WINE GRAPE VARIETIES

George Kerridge
and Allan Antcliff

REVISED EDITION

CSIRO
PUBLISHING

CSIRO Cataloguing-in-Publication Entry

Kerridge, George.
  Wine grape varieties.

  Bibliography.
  ISBN 0 643 05982 2.

  1. Grapes – Varieties – Australia. 2. Grapes – Varieties.
  I. Antcliff, A. J. II. Title.

634.830994

© CSIRO 1999

This edition previously published as *Wine Grape Varieties of Australia*
This edition first published 1999
Reprinted 2000, 2004

This book is available from:
CSIRO PUBLISHING
PO Box 1139
(150 Oxford Street)
Collingwood VIC 3066
Australia

Tel: (03) 9662 7666   Int: +61 3 9662 7666
Fax: (03) 9662 7555   Int: +61 3 9662 7555
Email: sales@publish.csiro.au
Web: www.publish.csiro.au

Contents

Preface vii
Acknowledgment x
Introduction 1
Wine grape varieties 6
Wine grape varieties new to this edition 183
Synonyms 191
Variety collection 195
Glossary 201
References 205

Preface

Wine Grape Varieties is based on four previous booklets on wine grape varieties published by CSIRO and written by the late Dr A.J. Antcliff, a Senior Principal Research Scientist at the then CSIRO Division of Horticultural Research, Merbein, Victoria. These are:

Some Wine Grape Varieties of Australia (Antcliff 1976),
Major Wine Grape Varieties of Australia (Antcliff 1979),
Minor Wine Grape Varieties of Australia (Antcliff 1983), and
Wine Grape Varieties of Australia (Kerridge and Antcliff 1996).

The present edition incorporates revised statistics and information on the varieties covered in the original books, plus information on additional varieties. The photographs, taken by Mr E.A. Lawton, are of material from vines growing at the CSIRO Division of Plant Industry (Horticulture Section), Merbein, or, in a few cases, at the Sunraysia Horticultural Research Institute of the Victorian Department of Agriculture, Irymple. In some cases the material was established with cuttings from old vineyards. We would like to acknowledge all owners who allowed inspection of their vineyards and special mention should be made of Bests Great Western, where numerous lengthy inspections were made of an old planting in which nearly 60 varieties were found. The identity of all varieties illustrated (except Moschata Paradisa and Solvorino, which are names used only in Australia) was confirmed by M. Paul Truel, INRA, Domaine de Vassal, France, during a visit to Merbein in February 1982.

Useful information for recognising the varieties is included, however, full descriptions have been avoided. These are available in overseas publications for the varieties as grown in those areas. Under different conditions of soil or climate, some characters of the vine may change and descriptions based on limited numbers of experimental vines might prove to be misleading. For similar reasons, the indications of cultural characteristics and potential yields that have been given where possible should not be taken too literally for any particular situation in Australia.

Vine identification – ampelography, from the Greek *ampelos* (vine) and *graphe* (description) – uses a phenotypical classification which permits the construction of a coherent system applicable to all species of the genus *Vitis*, their hybrids as well as varieties (Galet 1985). Ampelography is based on the morphological characters of the growing shoot tips, shoots and leaves as well as those of the bunches and berries. The most important factors considered are the hairiness of the growing tip and the young shoot, the hairiness of the adult leaf, and the leaf and cane shape.

Wine Grape Varieties illustrates fruiting canes, leaves and actively growing shoot tips. All illustrations are studio photographs thus, for comparisons, detached leaves, for example, should be similarly laid out on a flat surface. In the photographs of the detached leaves, the upper and lower surfaces of the leaves are shown separately, with the upper surface at the top. Where possible the detached leaves as well as the shoot tips have been photographed on a black background to give a better indication of the hairiness of the lower surface.

Sizes of leaves and berries can be greatly affected by weather conditions and only a rough indication is given of what can be expected under favourable conditions. Medium as applied to leaves means an average diameter of about 14–18 cm, and as applied to berries an average weight of about 2–4 g. A medium bunch would be about 15–20 cm long.

It is now possible with molecular biology techniques to establish DNA profiles or genetic 'fingerprints' which provide objective identification of varieties – even if two cultivars look identical, or one cultivar looks different when grown in another environment. The CSIRO Division of Plant Industry (Horticulture Section) has developed semi-automated DNA-profiling technology specifically for the identification of grapevine cultivars. Cultivars which have been profiled are identified in this book by the presence of the DNA logo **DNA**. DNA profiling is relatively expensive, however, and it will still be necessary to use the standard ampelography techniques to confirm the identity of large areas of vineyard.

The statistics for area are those collected by official organisations in each country. Several diseases are mentioned within the text. These are black spot (*Elsinoe ampelina*), downy mildew (*Plasmopara viticola*) and powdery mildew (*Uncinula necator*). 'Noble rot' is induced by *Botrytis cinerea*.

In the countries of the European Union, the term 'recommended' applied to a vine variety has a more defined connotation than in Australia or California. In these countries vines can be grown only on land with planting rights – the land may be used in full for recommended varieties, only in part for authorised varieties and not at all for others. The recommendations apply to local government areas and not the country as a whole.

George Kerridge

Acknowledgment

This book would not have been possible without the valuable input of Ms Angela Brennan, formerly Information Manager, CSIRO Division of Plant Industry (Horticulture Section). Angela was responsible for the collation of the photographs for this book and for providing much needed motivation during its formative stages. Without her this book might never have been completed.

Introduction

Grapevines were brought to Australia with the first fleet in 1788. Grape cuttings and seeds were collected in Rio de Janiero and from the Cape of Good Hope and planted at Port Jackson in Farm Cove, the site of the present Sydney Royal Botanic Gardens. In 1791 Governor Phillip established the first vineyard when he planted 1.2 ha of vines at Parramatta. Unfortunately there was very little knowledge of grapegrowing amongst the convicts and soldiers and in 1801 the Duke of Norfolk sent out two Frenchmen, Landrier and de Riveau, who had been prisoners-of-war held at Portsmouth. They had little success in controlling a major outbreak of 'blight' and were subsequently sent home (Gregory 1988).

In 1816, G. Blaxland planted a vineyard at Parramatta with vines introduced from the Cape of Good Hope. Wine from this vineyard was sent to London in 1822, where it was awarded a silver medal. A subsequent parcel of wine was awarded a gold medal in 1827. John Macarthur planted a vineyard at Camden Park in 1820 and by 1827 produced a vintage of 90 000 litres.

Interest in viticulture in the colony increased rapidly and in 1831 James Busby travelled through Spain and France collecting cuttings of grape varieties for the colony. He was recorded as having collected 433 varieties from the Botanic Gardens in Montpellier, 110 from the Luxembourg Gardens in Paris, 44 from Sion House near Kew Gardens in England and 91 from other parts of Spain and France. At this time, varieties were not well characterised and it seems certain that some were repeated in this introduction under more than one name, perhaps many more – the same name may also have been used for more than one variety. It is clear from the catalogue of the collection put out by the Sydney Botanic Gardens in 1842 that some of the varieties may also have been confused, for example, Semillon is described as a black grape and Malbec as a white. Unfortunately, this collection was removed in 1857 – but not before cuttings had been distributed to Camden, the Hunter Valley and the Adelaide Botanic Gardens from where they spread throughout South Australia.

While the original collection and those established from it have been lost, more of the varieties have survived in Australia than is generally realised. From the localities in which they have been subsequently found, it seems very likely that there are vines of varieties such as Crouchen, Chenin Blanc and Ondenc, as well as better known varieties such as Semillon, Riesling, Shiraz and Cabernet Sauvignon which can be traced back to Busby (even though the major plantings of some of these varieties may have come from other sources). Among the minor varieties, the discovery of surviving vines of Bourboulenc, Piquepoul Noir, Tocai Friulano and Troyen was of great interest. Other varieties since found and identified were Fer, Gamay, Gueche and Pougnet, and about 20 more varieties have been distinguished but not yet identified. There are also varieties from older collections with obviously local names which remain to be identified.

Vineyards rapidly spread to the rest of the Australian colonies – vineyards were planted in the Yarra Valley in Victoria in 1830 and Adelaide in 1837. The first vineyard in the Barossa Valley in South Australia was planted by Johann Gramp at Jacob's Creek in 1847. The first Western Australian plantings were made on the Swan River near Perth in 1829, and the first plantings in Queensland were at Stanthorpe in 1859 and at Roma in 1863.

The introduction of grape *phylloxera, Daktulosphaira vitifolii*, first reported at Fyansford near Geelong in Victoria in 1875, devastated the industry and necessitated the costly process of replanting the infected areas with vines grafted onto resistant rootstocks. Strict quarantine regulations have restricted the spread of this serious pest and today most of Australia's vineyards are free of phylloxera.

INTRODUCTION

The arrival of the Chaffey brothers from California in 1886 saw the expansion of the irrigated horticultural regions near Mildura in Victoria and Renmark in South Australia. Further irrigated areas were developed in NSW with the settlement of the Murrumbidgee Irrigation Area commencing in 1912. These three regions now produce approximately 75% of Australia's winegrapes. These regions initially included major plantings of the classic table wine varieties, but in the early 1900s these were almost all replaced by fortified wine varieties to supply the United Kingdom market. This situation continued until the 1950s when an increasingly multicultural Australian population began demanding high quality table wines and the varietal mix again swung in favour of the classic wine varieties such as Chardonnay and Cabernet Sauvignon. Australia was fortunate in having large areas of Shiraz, originally planted for port wine styles, which, in the warm Australian climate, has proved to be ideally suited for the production of full-bodied red wine styles .

There has been much confusion of names for grape varieties in Australia. Names were often misspelt or exchanged on the early introductions, and even relatively recent imports from California and elsewhere were seen to be incorrect. Dr Antcliff began collecting varieties into a major germplasm collection at the CSIRO Division of Horticulture at Merbein, Victoria in the late 1950s for use in a grape-breeding program which commenced in the early 1960s. It soon became apparent that many of the varieties were wrongly named and in 1976 M. Paul Truel was brought to Australia to sort out the confusion (Antcliff 1976).

M. Truel was the curator of the INRA grape germplasm collection at Vassal near Montpellier in the south of France. This collection contains over 3700 grape varieties with approximately 5600 clones of *Vitis vinifera* and 1700 species hybrids, either rootstocks or direct producers.

Five plants of each clone are maintained at Vassal. As well, representative leaves are dried and stored in a herbarium and photographs of leaves, flowers, bunches and berries are taken to allow comparisons at any time of the year. Standard descriptions have been prepared and a data-retrieval system established in which the clones are classified for a range of characters, e.g. five classes for leaf lobing and 10 classes for berry shape. These aids can suggest relationships, but in the final analysis it is the examination of the vines in the field at several times during the growing season which allows clones to be identified without ambiguity as belonging to the same or different varieties.

During his visit to Australia, M. Truel found that most of the major wine grape varieties – Shiraz, Grenache, Cabernet Sauvignon, Mataro, Riesling, Semillon, Doradillo, Pedro Ximenez, Trebbiano and Palomino – were generally correctly identified, although some incorrectly named plantings were found (e.g. a Riesling called Semillon). Some of the minor varieties – Graciano, Marsanne, Mondeuse and Sauvignon Blanc – were also found to be correctly identified. However, some of the errors included Chenin Blanc called Semillon in Western Australia and Albillo in South Australia. Chenin Blanc was also the major component of a Chardonnay planting in the Rutherglen area of Victoria. Bastardo was found under the name of Touriga in NSW and as Cabernet Gros in South Australia. The true Malbec was found to be correctly identified, but a Malbec in South Australia was found to be Tinta Amarella and one in Victoria was found to be Dolcetto. More recent introductions were also found to be incorrect, e.g. a Gamay Beaujolais which was actually a clone of Pinot Noir, and Napa Gamay which proved to be Valdiguie. Even more recently, two clones of Pinot Blanc were found to be Semillon.

In the early 1980s the OIV (International Office for Grapes and Wine), UPOV (International Union for the Protection of New Varieties of Plants) and the FAO (Food and Agriculture Organisation of the United Nations) decided on a uniform approach to the description of the morphological characteristics of grapevine cultivars. This resulted in the OIV Descriptor List for Grapevine varieties and *Vitis* species – a list containing descriptors for 128 characteristics including morphological descriptors for shoot, inflorescence, leaf, bunch and berries. Secondary information on seeds and dormant canes increased the accuracy of this system.

During the 5th International Symposium on Grape Breeding held in Germany in 1989, a new 'Preliminary Minimal Descriptor List for Grapevine Varieties' was proposed. This list included 41 characteristics of which 12 descriptors were taken from the OIV Descriptor List for Grapevine Varieties and *Vitis* species without any adjustment, 19 descriptors were modified and 10 new descriptors were added (E. Dettweiler 1991).

To date, 35 collection holders from 20 countries have participated in the description of cultivars. They recorded 17 visual and 3 berry descriptors and sent leaf and seed samples for verification at the Institute of Grape Breeding, Geilweilerhof, Germany. Leaf comparison of cultivars with the same name but from different sites, and comparison with illustrations revealed that 98% of cultivars were kept under the correct name or a valid synonym. This Preliminary Minimal Descriptor List will probably become the standard rapid ampelography system of the future, although the full 128 character list may still be required for registration of new varieties with the OIV and EU.

Viala and Vermorel in their seven volume *Ampelography*, published in 1909, list 24 000 names for 5200 varieties. Alleweldt and Dettweiler (1992) in their publication, *The Genetic Resources of* Vitis, suggest that the more than 40 000 cultivars, rootstocks, strains etc. existing worldwide in collections, and the more than 30 000 cultivars, rootstocks and *Vitis* species described in the literature could be reduced to a little more than 15 000 genotypes (prime names of cultivars), including approximately 75 *Vitis* species.

The grape germplasm collection at the CSIRO Division of Plant Industry (Horticulture Section) at Merbein, Victoria, contains almost 700 grape varieties or species. This book describes 92 varieties from the Merbein collection. It includes all of the varieties commonly grown in Australia plus some that could have a place in the future. It will be a useful guide for the practising viticulturist, students of viticulture and for the enthusiastic amateur.

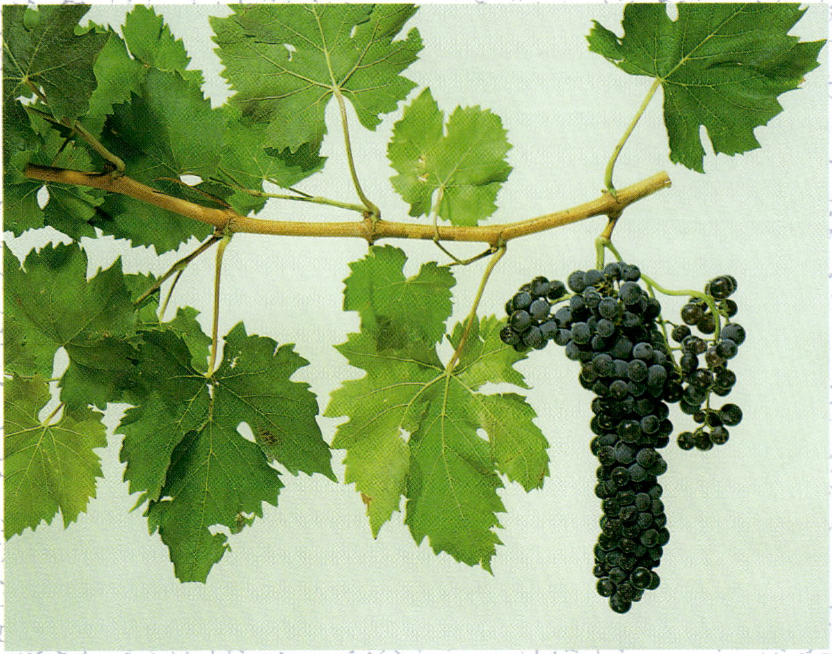

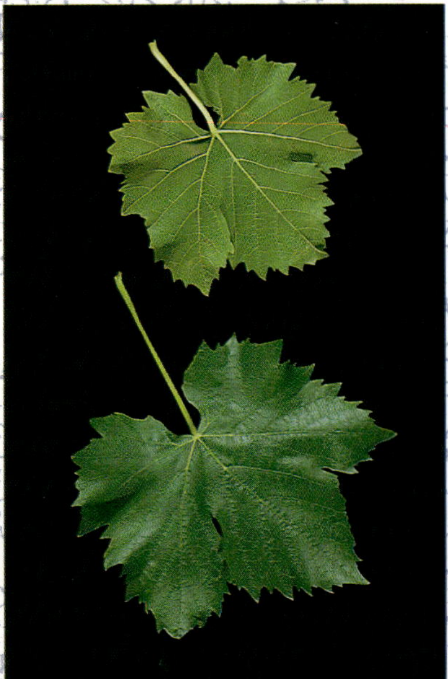

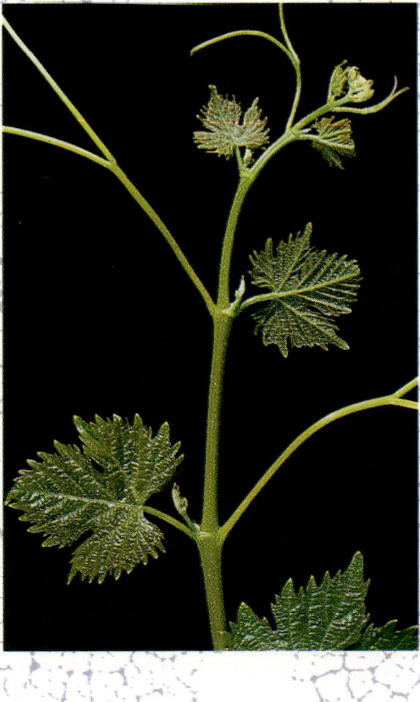

Aleatico

Aleatico is an Italian variety which can be found in most areas of the country, with the most substantial plantings being in Tuscany and the island of Elba. It is also grown in Corsica and it is a minor variety in California with about 40 ha planted.

Aleatico is a vigorous variety with an upright habit of growth. It has medium, 3- to 5-lobed, plane and slightly rough leaves which are hairless above and below. The petiolar sinus is in the shape of a narrow lyre. Bunches are medium, more or less cylindrical and well filled to compact with medium, round berries with a heavy bloom and a muscat flavour. Although it is a different variety it is somewhat like a black form of Muscat à petits grains.

In Italy, Aleatico is used to make a sweet, red, muscat wine which is highly regarded. An unfortified sweet muscat wine with a bright, light red colour would be unusual in Australia, and could be a marketable product. However, Aleatico can be used to make white wines and fortified wines and could be tried as a substitute for Muscat à petits grains where there were problems with that variety.

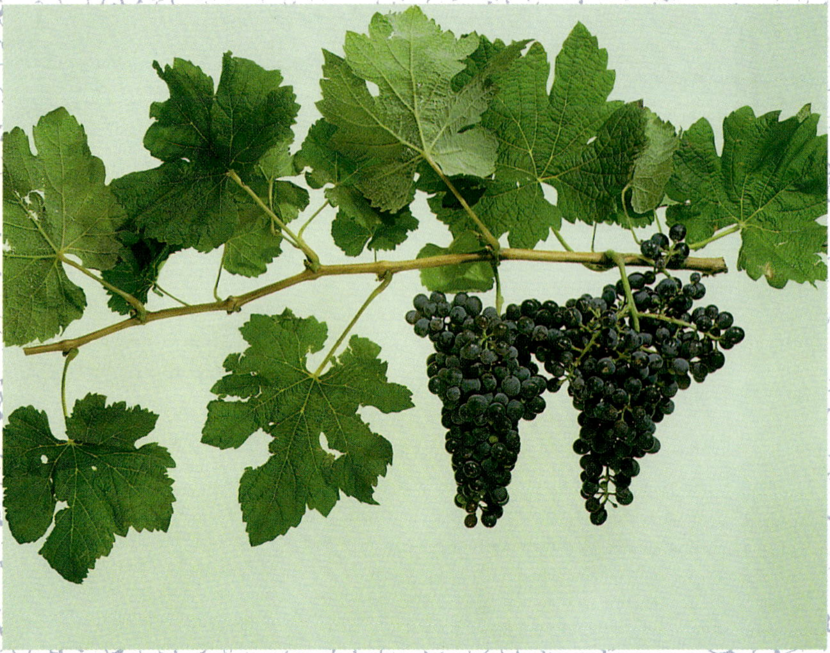

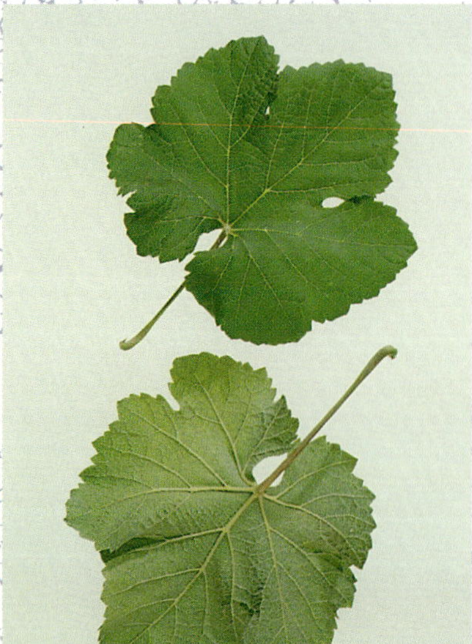

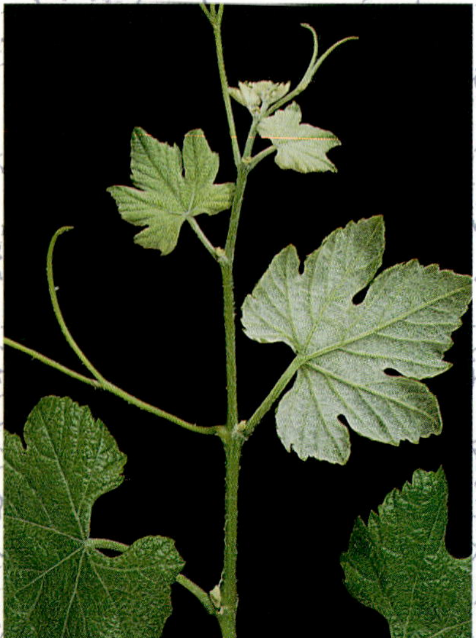

Alvarelhao

lvarelhao was recommended with Bastardo and Touriga by Mr F. de Castella, former Government Viticulturist in Victoria, for the production of port. However, in the variety classification used in the Douro Valley in Portugal, while Bastardo and Touriga are rated very good, Alvarelhao is rated as only reasonable. Thus it is perhaps not surprising that less Alvarelhao than Touriga and Bastardo has been planted in Australia, there being a few small plantings in north-east Victoria and southern New South Wales only. There appears to be very little of this variety grown outside Portugal. It is not clear whether it is present in California as the variety imported from there as Alvarelhao proved to be Touriga.

Alvarelhao is a vigorous variety with a fairly upright habit of growth. The leaves are medium to large, rough, dark green, 3- to 5-lobed tending to fold inwards about the midrib and roll back at the edges, with some tufted hairs on the lower surface and a V-shaped petiolar sinus. Bunches are medium, conical and rather loose, with small, short elliptical berries.

Alvarelhao is low in colour and does not appear to have any particular virtues for either port or dry red wines.

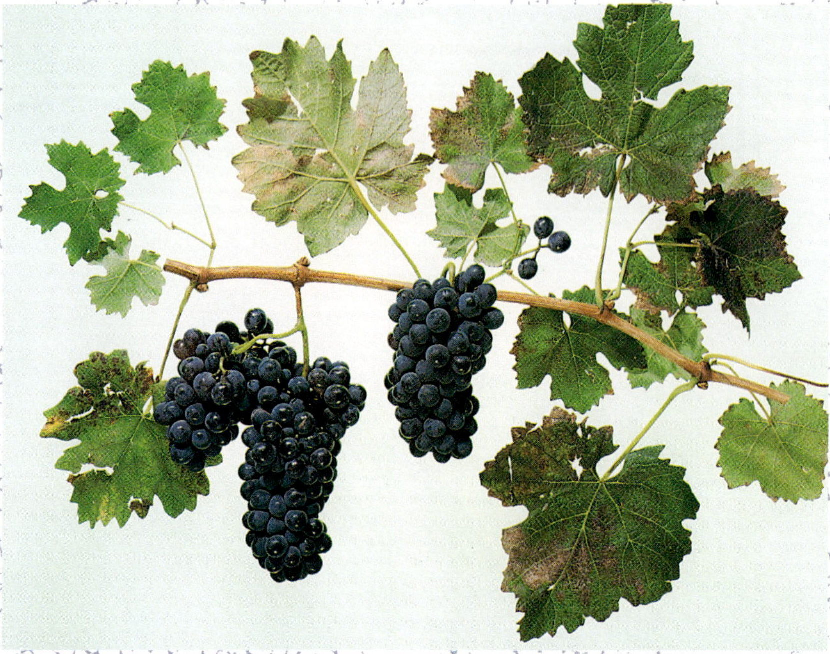

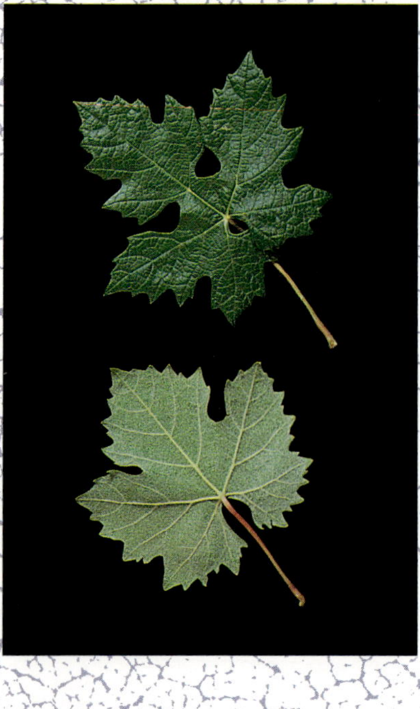

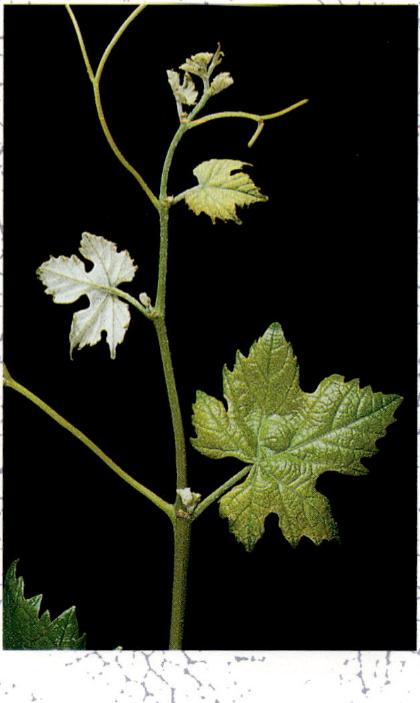

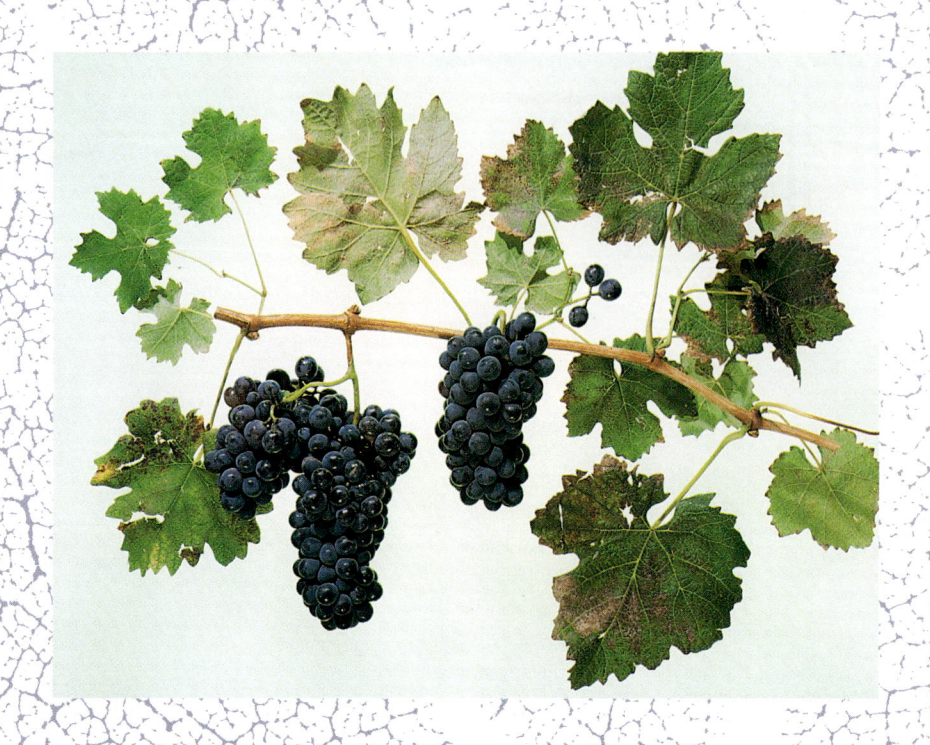

Barbera

Barbera is the leading wine grape of Italy, grown mainly in Piedmont. Because of mixed plantings the area is uncertain but the annual production is of the order of half a million tonnes. It is used in wines of controlled appellation, sometimes alone and sometimes mixed with other varieties. In Argentina there are about 1000 ha of Barbera. Plantings in California increased sharply from about 1970 reaching 8600 ha in 1977 and declining to 4100 ha by 1992. Only a few small plantings have been made in Australia.

Barbera is a reasonably vigorous variety although its rather slender shoots and open foliage tend to give it a sparse appearance. The leaves, which are easily damaged by wind, are of medium size, plane, slightly rough, rather deeply 5-lobed and hairy on the lower surface. The petiolar sinus is of a narrow lyre shape, often closed and sometimes with partly overlapping edges. Bunches are up to medium in size, conical to irregular in shape, sometimes winged and well filled to compact. Berries are medium, oval and intensely coloured with a heavy bloom. The long, green bunch stalks make it easy to harvest by hand and it is one of the most satisfactory varieties for mechanical harvesting.

Wines from Barbera should have good colour, tannin and acidity, and a distinctive varietal character which, being unfamiliar, may not be immediately acceptable in Australia. In Italy it is mostly used for dry red wines, but sweet red and sparkling red wines are also made.

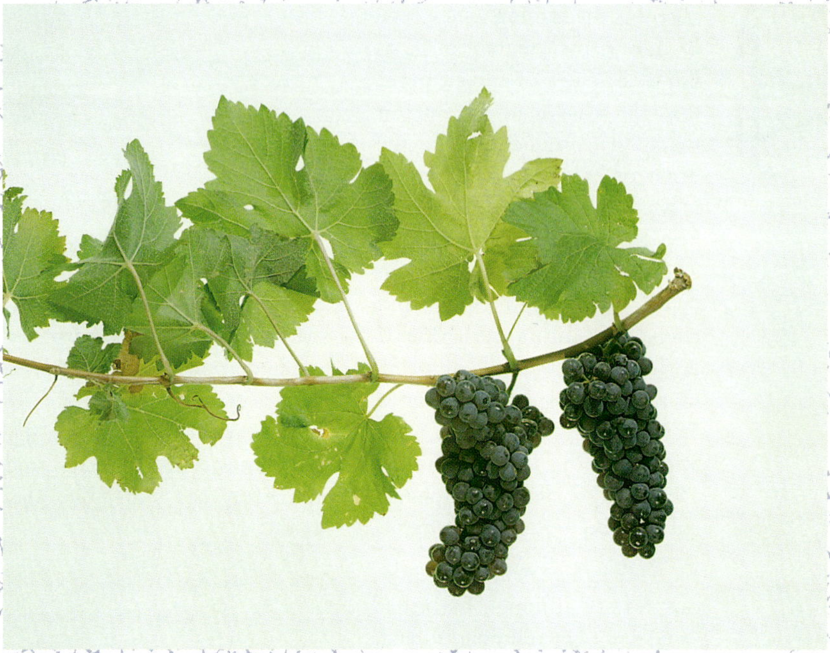
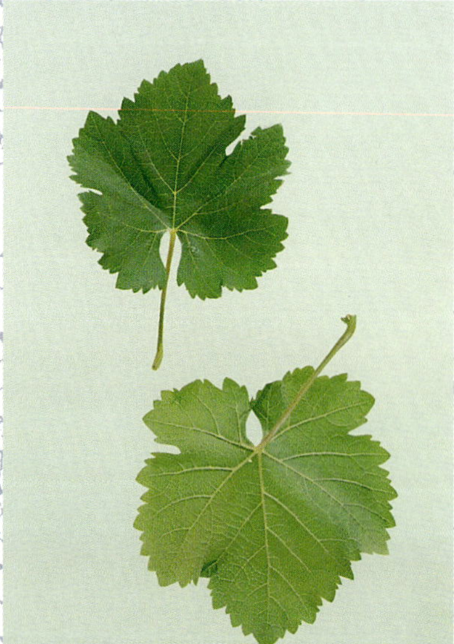
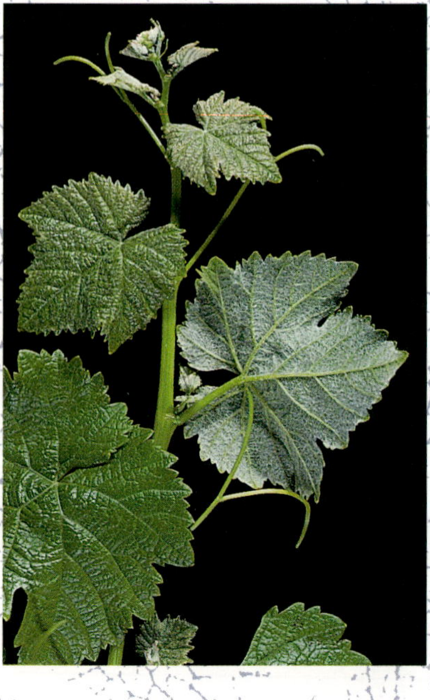

Bastardo

Bastardo is considered one of the better port varieties in Portugal although it is not as widely grown as Touriga or Tinta Amarella. It is also grown, to the extent of 140 ha, under the name of Trousseau, in the Jura region in the east of France. It is probably one of the 'port sorts' of which 188 ha are recorded in South Africa and there may be a little in California and South America, but not enough to be recorded separately. The exact area in Australia is uncertain. There are 26 ha in South Australia as Cabernet Gros, a little in north east Victoria and nearby in New South Wales called Bastardo, and some plantings called Touriga in New South Wales are also Bastardo.

Bastardo is a fairly vigorous variety with a spreading habit of growth. It has medium to rather small, lightish green leaves, entire to slightly 3-lobed, smooth above with a few cobwebby hairs below. The petiolar sinus is more or less V-shaped, often narrow with almost parallel sides. The bunches are rather small, generally cylindrical, with small, round, soft berries with a heavy bloom. The fruit ripens early and attains a high sugar concentration which increases even further as the berries wilt.

Under most Australian conditions Bastardo is best suited for fortified wines. It does not provide much colour in the wine but will combine with other varieties which can provide colour and flavour.

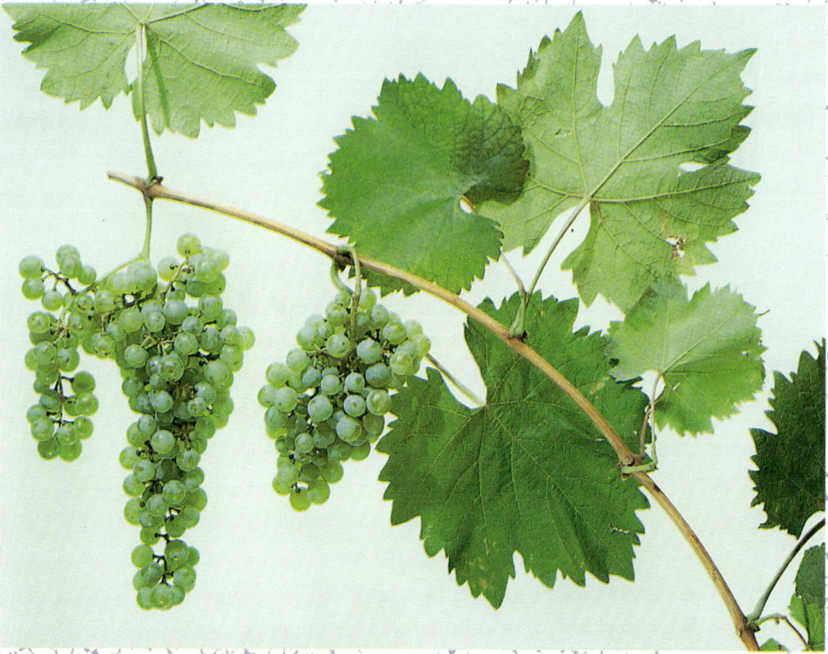

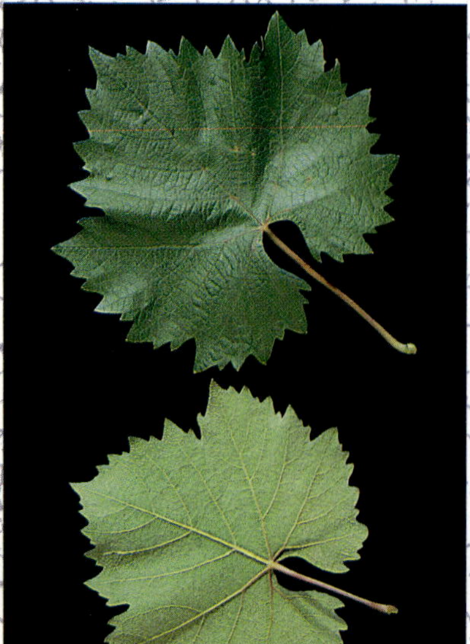

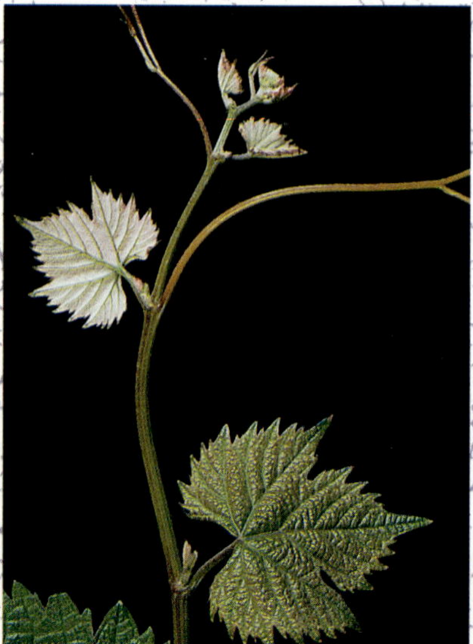

Bianco d'Alessano

Bianco d'Alessano is a late ripening white wine grape variety from the Puglia region of southeast Italy. There are 11 000 ha of this variety planted in the province of Taranto near Bari.

It is a vigorous variety which was one of the latest ripening of all of the white wine varieties in the grape germplasm collection at the CSIRO Division of Horticulture at Merbein. The newly opened buds are large and downy with carmine colouration. The young shoot tips are white in appearance with carmine edges on the young leaves. Adult leaves are large, orbicular, 3- to 5-lobed, glabrous, dull green on the upper surface and covered with felted down on the lower. The petiolar sinus is a narrow lyre and the leaf is lightly undulating. The teeth are irregular in size and convex on a wide base. The petiole is medium in length, hairy and rose coloured. The bunch is conical/cylindrical, compact, large and winged. The berry is greenish white, round, regular and of medium size. The juice is sweet and neutral in flavour. Canes are medium in length, uniform light brown in colour and striated.

Bianco d'Alessano produces yields of about 30 tonnes per hectare in the warm irrigated regions of Australia. The wines however tend to lack character and have been given only average scores by tasting panels.

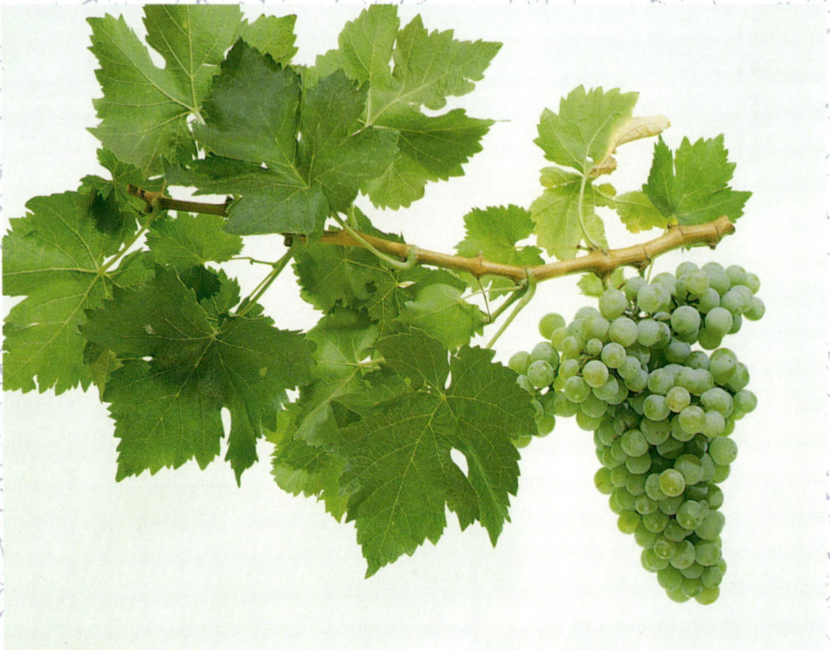
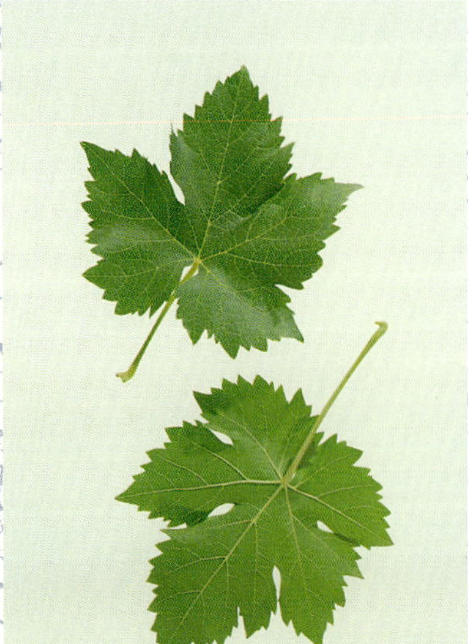
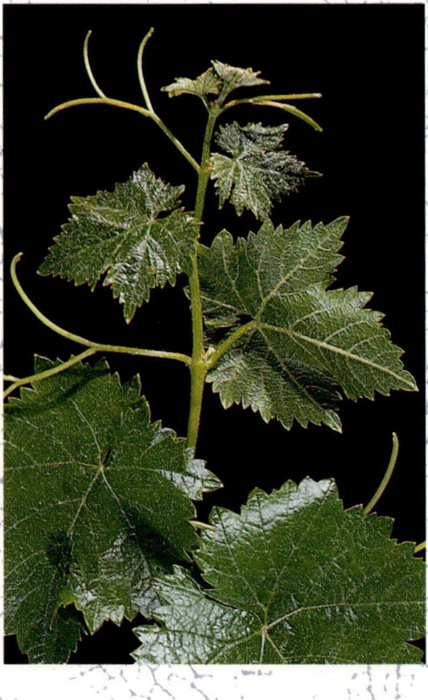

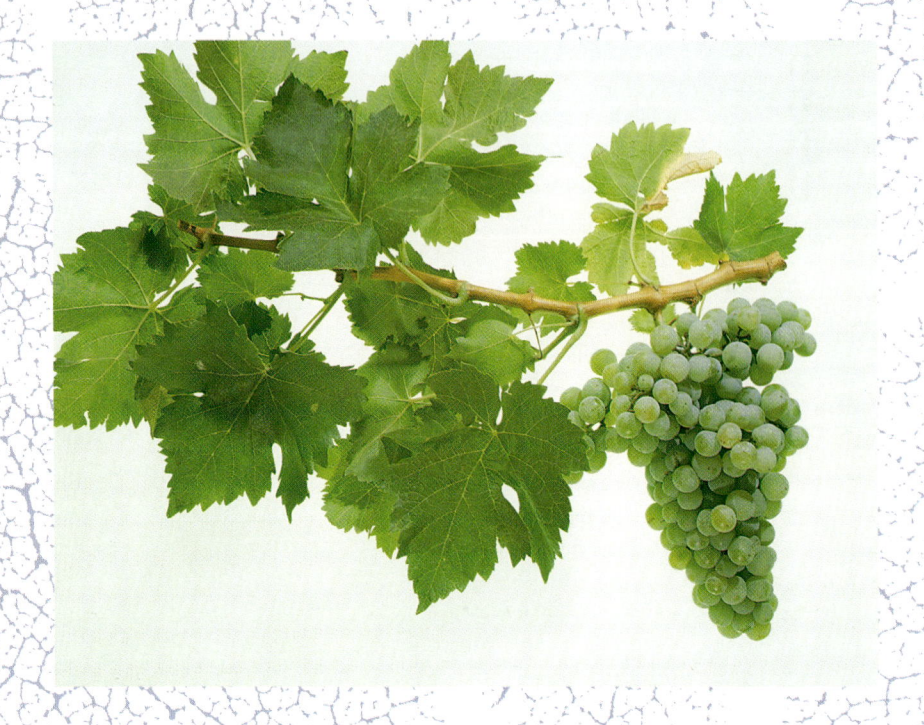

Biancone

Biancone has the distinction of giving the highest yield commercially of any variety in Australia. There are only 57 ha, almost all in the Riverland area of South Australia, but the average yield regularly approaches 30 tonnes per hectare, nearly half as much again as that of the next highest yielding variety, Doradillo, under similar conditions. The variety comes from Corsica, where its excellent production has led to one of its names being Pagadebiti, literally 'payer of debts'. Apart from isolated vines known as Green Doradillo or Late Doradillo the variety has been called White Grenache in Australia. Although the Grenache Blanc of the south of France, of which more than 12 000 ha have been planted, is the white form of the true Grenache, there have also been small areas of Biancone grown under the name of Grenache Blanc Productif. It was probably imported into Australia under this name. The Biancone of the Isle of Elba is thought to be the same variety but it does not appear to have become established in any other countries.

Biancone is a vigorous and rather upright growing variety. It has medium, clear green leaves, usually deeply 5-lobed, which in the Murray Valley are hairless on the lower surface, although in Europe they are described as having cobwebby hairs. The bunches are medium to large, conical and well filled, with medium, round berries often flattened at the stalk end. The berries are firmly attached and have a soft texture but tough skin.

In the Riverland, Biancone would be regarded as a variety for distillation or the production of bulk wine but there is some evidence that it can produce a distinctive dry white wine in cooler areas.

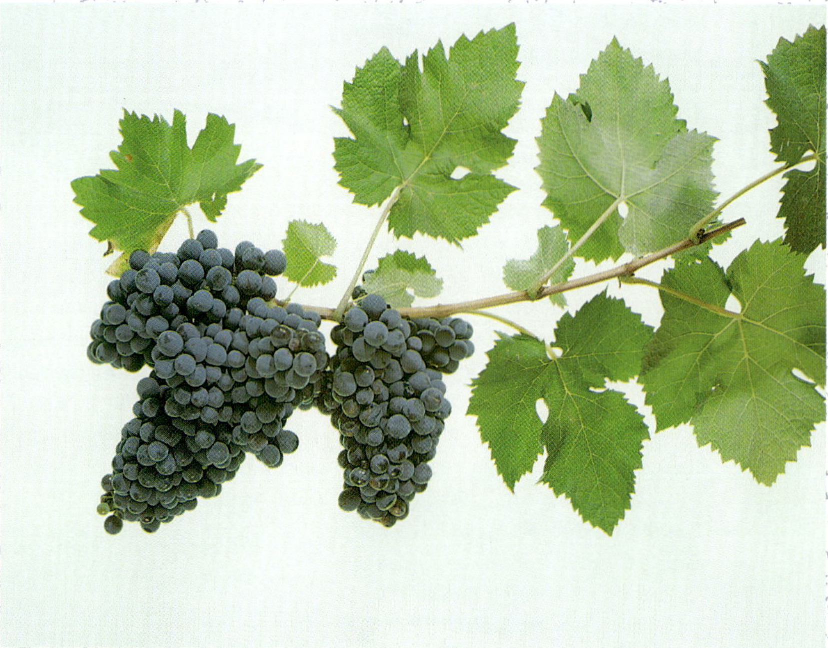

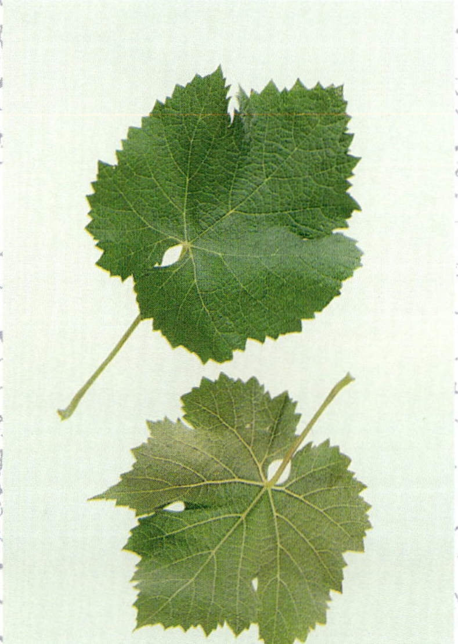

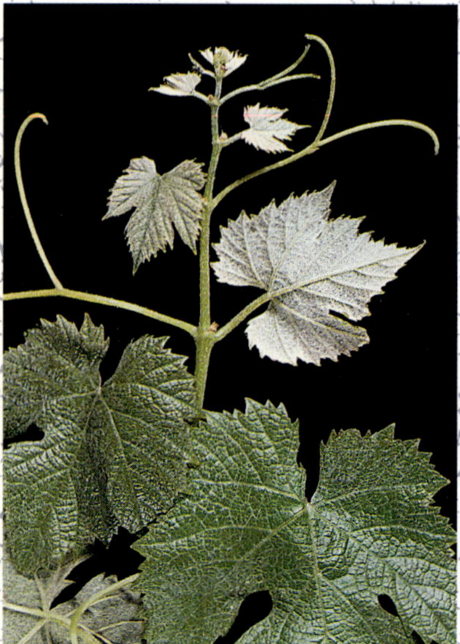

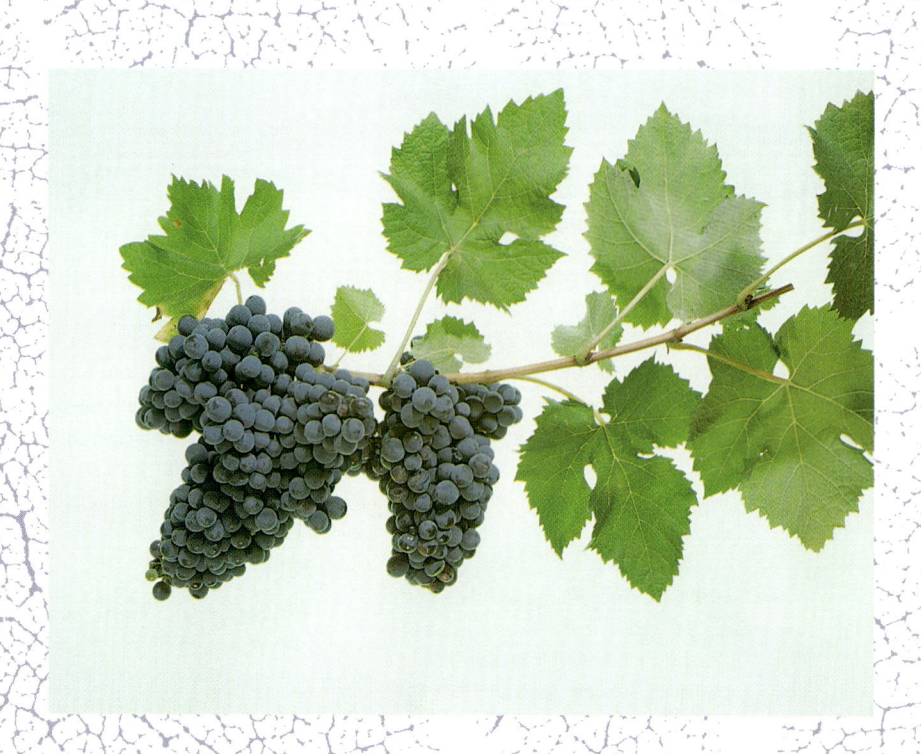

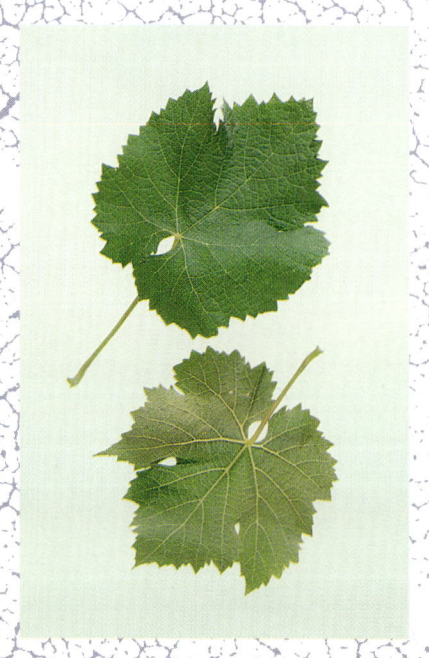

Bonvedro

Bonvedro is the Portuguese name of this variety, which is also grown in north eastern Spain as Cuatendra. It is possible that it also occurred in France as an obscure variety and perhaps came into Australia in a large collection such as that of Busby. In this way it could have become confused with Carignan, the name generally used for Bonvedro in Australia. There may also have been confusion with another variety from north-eastern Spain, Miguel de Arco, as the vines grown under this name in Australia also seem to be Bonvedro. In all there are about 57 ha of Bonvedro in Australia, mostly in South Australia with a little in New South Wales and Victoria.

Bonvedro is a vigorous and rather erect growing variety. It has medium to large, 3- to 5-lobed leaves which are rough and undulating, the undulation often causing the basal lobes to overlap leaving a 'peephole' in the petiolar sinus. The leaves are a somewhat greyish mid green, with dense white hairs on the lower surface. Bunches are medium, conical and compact with a woody stalk. Berries are medium, short oval, with a heavy bloom, soft flesh and a tough skin.

Wines made from Bonvedro in Australia, while having a pleasant varietal character, have suffered by comparison with wines from other varieties which have more colour and tannin. Increasing interest in lighter red wines may encourage a reappraisal of this position.

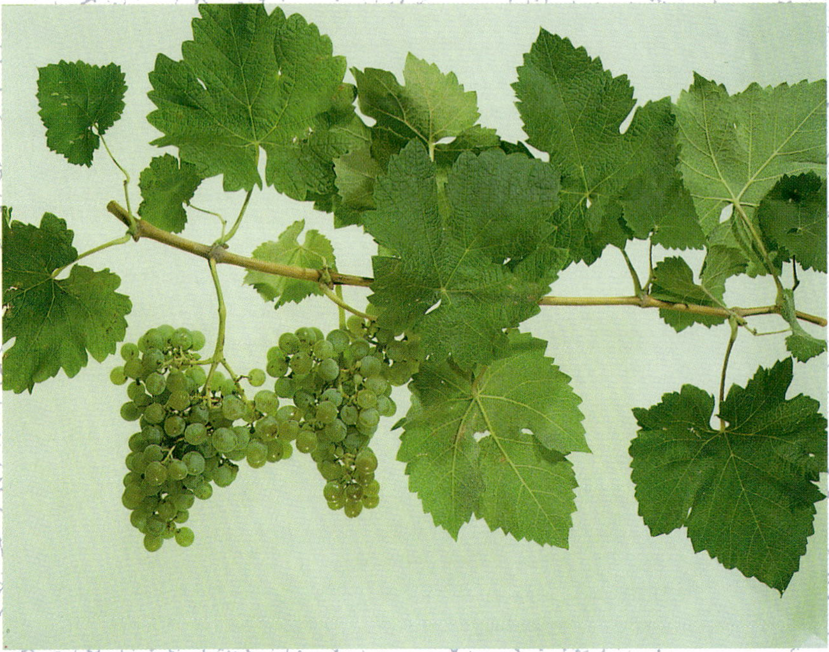

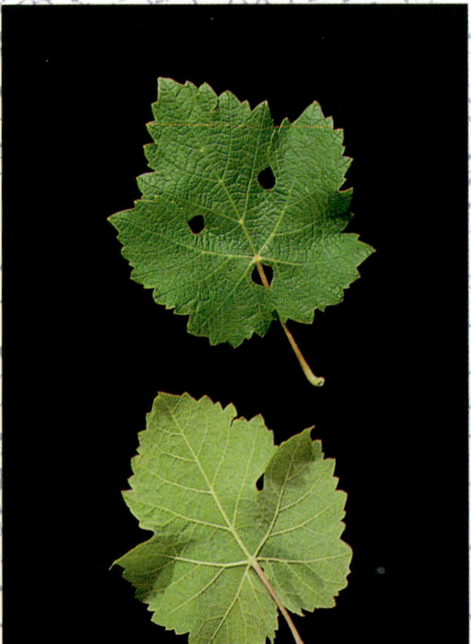

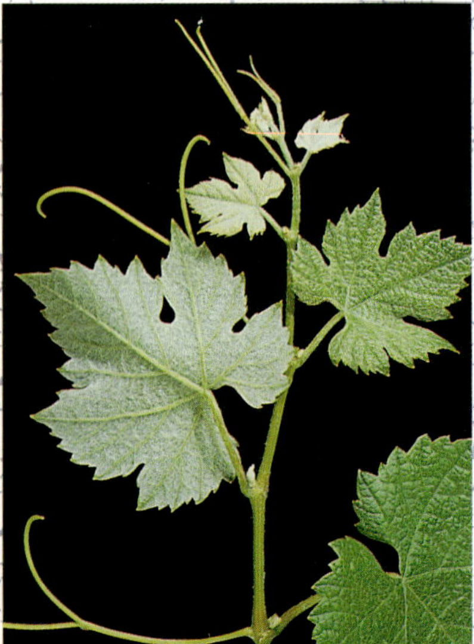

Bourboulenc

Bourboulenc is a recommended variety throughout the Mediterranean region of France and is found mainly in the lower valley of the Rhone. Although it is an approved variety for such wines of controlled appellation as Chateauneuf-du-Pape and Cotes du Rhone, the area of Bourboulenc in France declined from 703 ha to 383 ha between 1968 and 1979, but increased again to 723 ha by 1988. It appears four times, under different names, among Busby's imports into Australia in 1832, but has not survived in any of the official viticultural collections. It is now found only as odd vines in old vineyards in Great Western and Rutherglen and possibly elsewhere.

Bourboulenc is a vigorous variety with a rather spreading habit of growth. Leaves are medium, 3- to 5-lobed, thick, rough and sometimes undulating and downy white on the lower surface. The petiolar sinus is lyre shaped with the sides often overlapping at the top. Bunches are no more than medium in size, with long stalks, conical and well filled, with medium, oval berries. The resemblance to Clairette is indicated by the synonyms of Grosse Clairette and Clairette Doree, but the leaves of Bourboulenc are not quite as downy on the lower surface and the sides of the petiolar sinus do not overlap as far.

In France, standard dry white wines from Bourboulenc are thought to lack distinction although some varietal character is said to develop with bottle age. Wine from very ripe grapes has a more special character. In practice Bourboulenc is usually harvested and fermented mixed with other varieties.

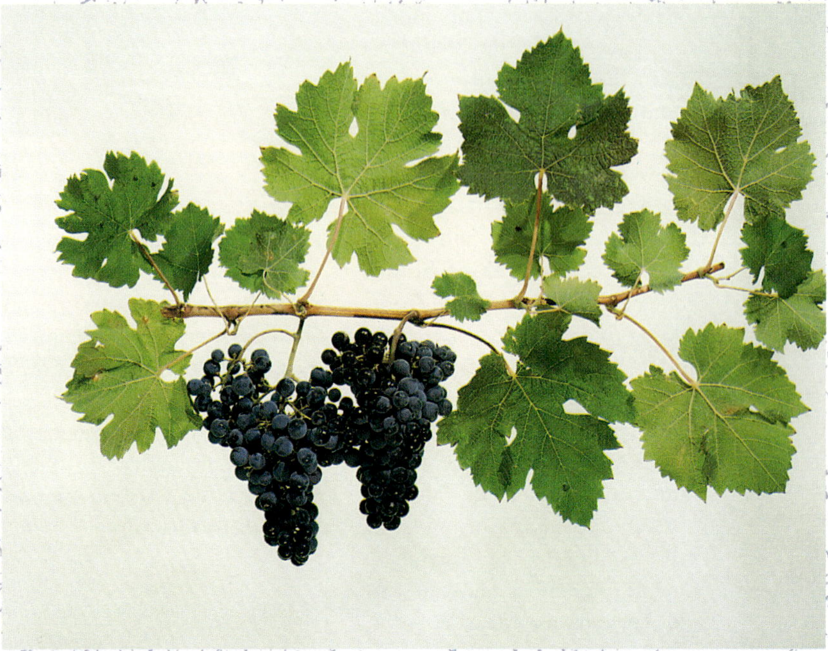

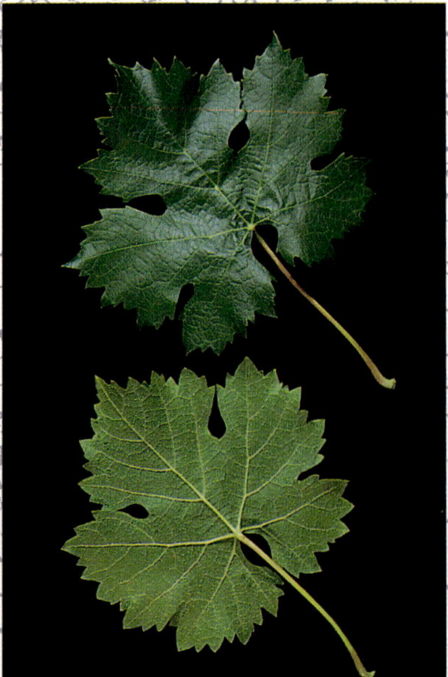

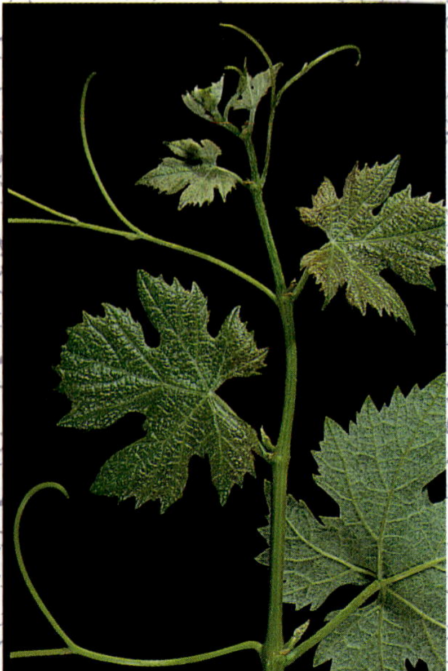

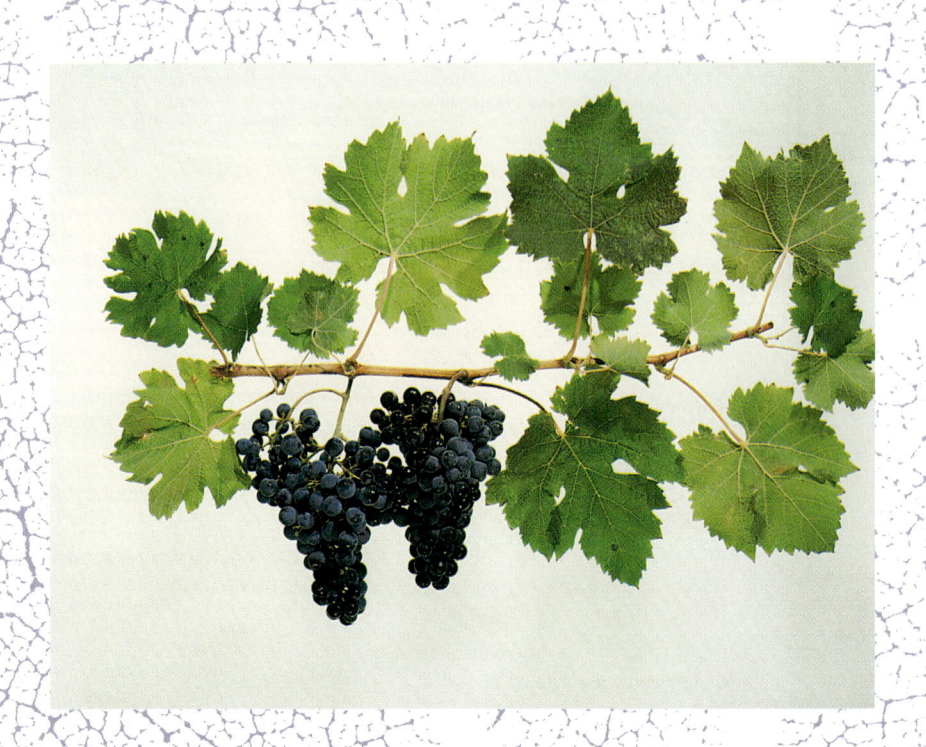

Cabernet Franc

Cabernet Franc is an important variety of the Bordeaux area. It should not be confused with so-called Cabernet Gros of South Australia, which is really Bastardo. Although some small plantings have been made in recent years, Cabernet Franc has generally been found in Australia as odd vines in plantings of Cabernet Sauvignon, particularly in north east Victoria where it sometimes occurs to the extent of more than one vine in ten. In France, where it is also grown in the Loire Valley and is now recommended throughout the entire south including Corsica, plantings increased from 9700 ha in 1958 to 30 256 ha in 1988. In Italy it is regarded more highly than Cabernet Sauvignon and recommended in more provinces. The Cabernet varieties are important in eastern Europe and South America, probably Cabernet Sauvignon the more so, but they are not always recorded separately. Cabernet Franc has not achieved the same recognition as Cabernet Sauvignon in California with 767 ha planted by 1992. The area planted to this variety in Australia is increasing with 453 ha planted by 1992.

Cabernet Franc is a vigorous and upright grower with medium, mid-green, rather rough, 3- to 5-lobed leaves with tufted hairs on the lower surface. The leaves are less deeply lobed than those of Cabernet Sauvignon and have the petiolar sinus in the shape of a narrow lyre. Bunches are small to medium, more or less cylindrical and rather loose with small, round berries. As with Cabernet Sauvignon there seem to be inferior vines which yield poorly, but a good clone held at Merbein has yielded at the rate of more than 15 tonnes per hectare.

The wine of Cabernet Franc has a pronounced varietal character, good tannin and colour, and ages well. It can be distinguished from the wine of Cabernet Sauvignon and the presence of Cabernet Franc may be partly responsible for the special character of Cabernet wines from north-east Victoria.

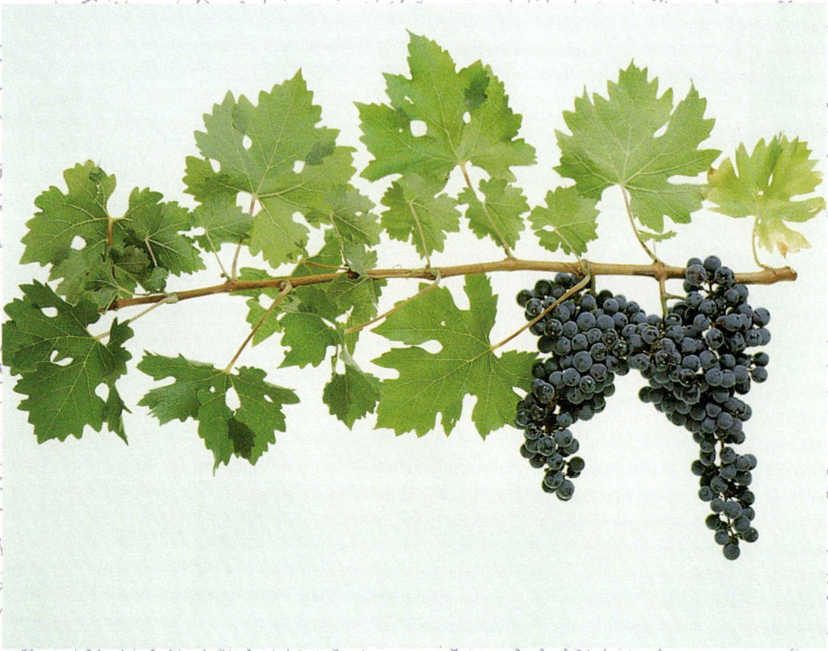

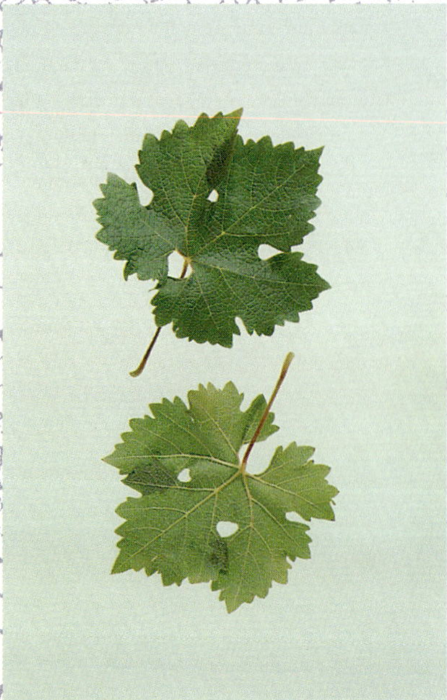

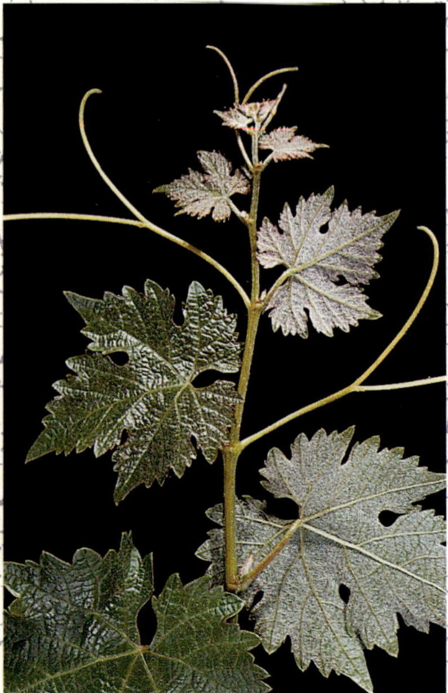

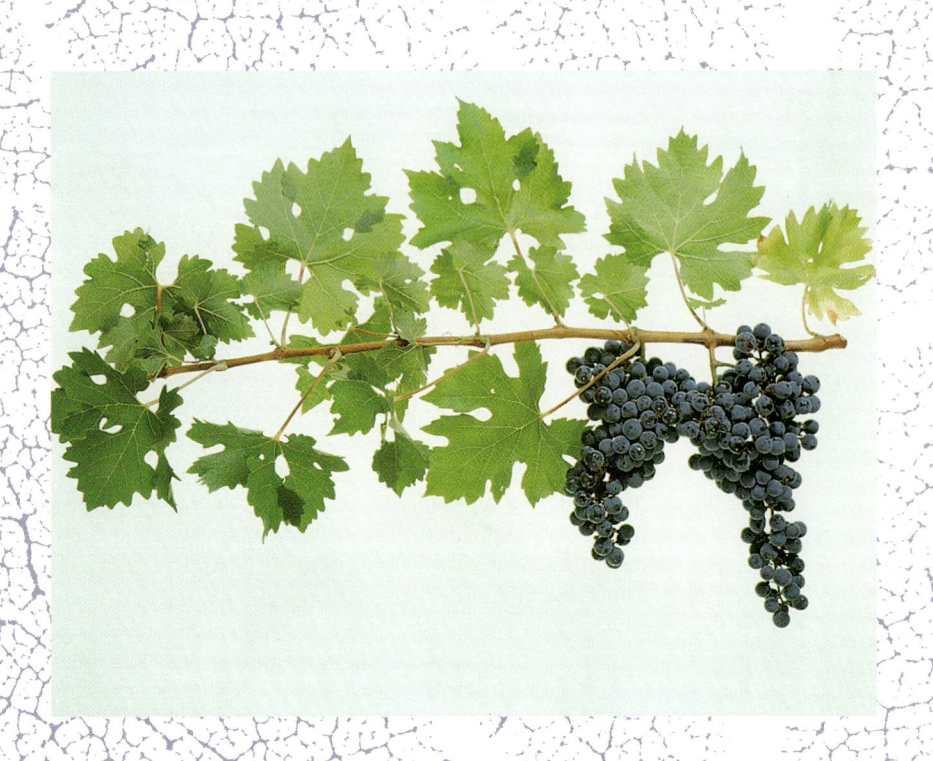

Cabernet Sauvignon

Cabernet Sauvignon comes from the Bordeaux region of France and it is the major variety in some of the best wines of the Medoc. The area in France increased from 11 800 ha in 1968 to 36 468 ha in 1988. It is also prominent in Chile with 12 500 ha and has increased rapidly in recent years to 14 000 ha in California, 5890 ha in Australia and 4660 ha in South Africa. In Italy it is a minor variety recommended only in the extreme north. It is probably more important in eastern Europe and Argentina.

Cabernet Sauvignon is vigorous with rather upright growth and is often cane pruned to avoid limiting the crop. It is generally regarded as low yielding, but with clonal selection it has been possible to combine good yield with good wine quality. It has medium deeply 5-lobed leaves with the petiolar sinus often cut right in to the veins at the base. The leaves are glossy green above and have scattered tufts of hair on the lower surface. The bunches are rather small, conical, often winged and loose to well filled. The small round berries detach cleanly for mechanical harvesting.

The excellent quality of the wines of Cabernet Sauvignon is well known. They have good colour and a pronounced varietal character which is very intense when the vines are grown under cooler conditions. With their high tannin they require aging and are often blended.

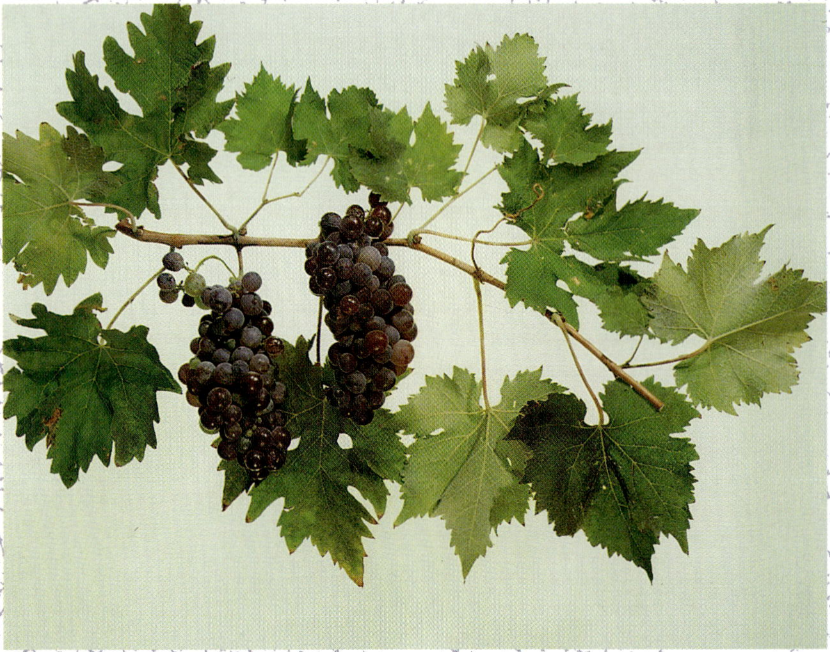

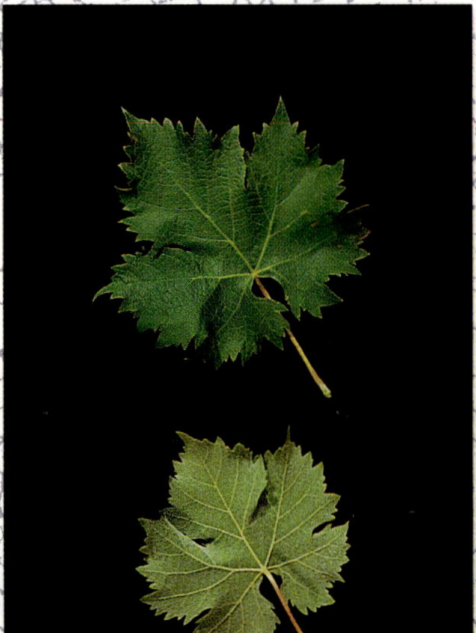

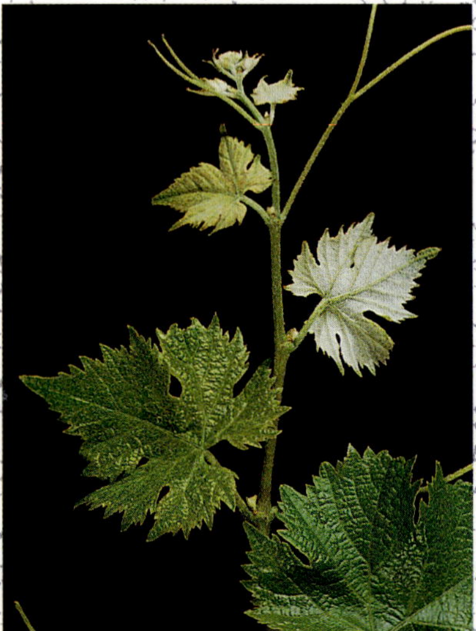

Calitor

Calitor is now a recommended variety only in the Departement of Var in Provence in France, and the area is being maintained at about 300 ha. It is approved as an accessory variety for the wine of appellation Bandol. Five of Busby's imports into Australia in 1832 appear to be Calitor under various synonyms, and it seems to have enjoyed some popularity for a time under the name of Pride of Australia. It is now found only in variety collections and as odd vines in old vineyards.

Calitor is a vigorous variety with an upright habit of growth. Leaves are medium to large, deeply 5-lobed with narrow pointed teeth, rough and undulating, sometimes puckered, and downy white on the lower surface. The petiolar sinus is a narrow lyre. Bunches are medium to large and compact. A characteristic feature is a pronounced bend in the bunch stalk at the node and the stalk is usually woody to this point. Berries are medium but irregular in size, round and rather poorly coloured.

Wine from Calitor alone is likely to be light in body and colour and Calitor would be used in conjunction with varieties rich in tannin such as Mataro and Carignan. It does not appear to offer any advantages over varieties more commonly used in Australia.

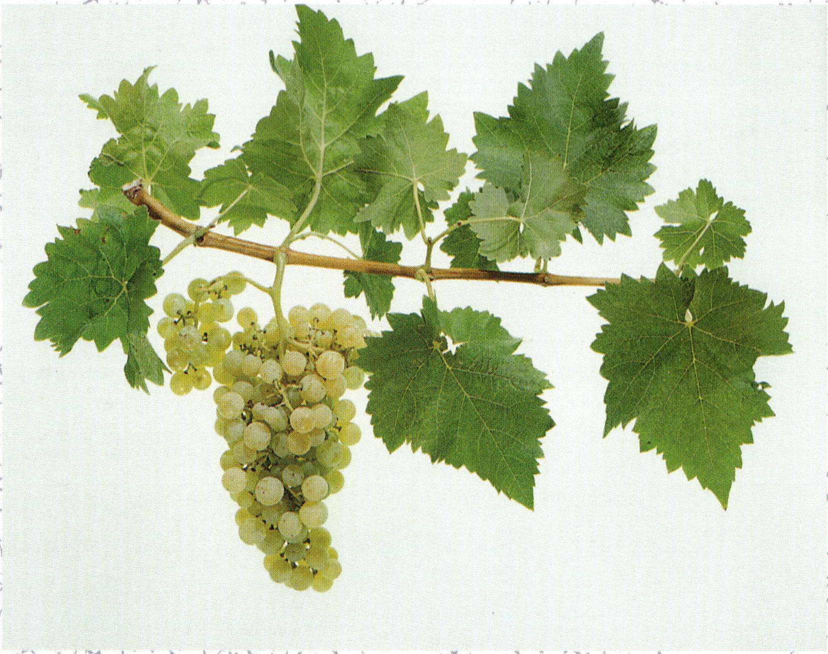

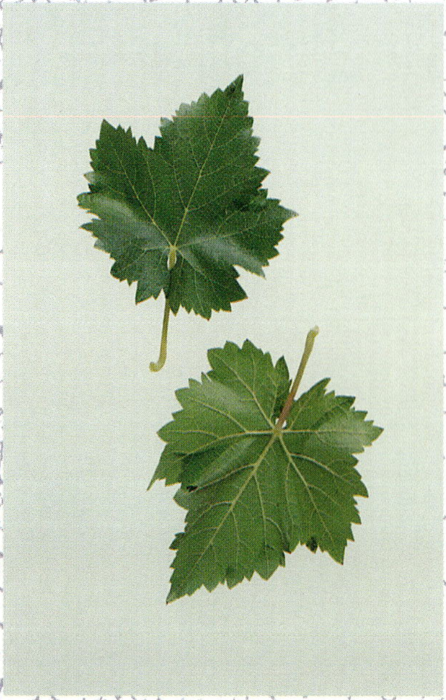

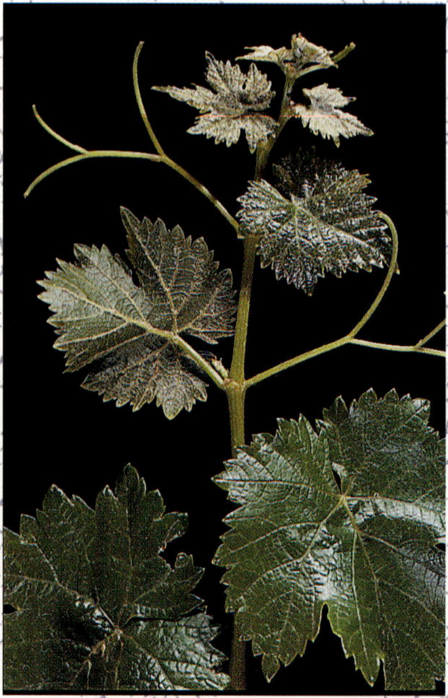

Cañocazo

Cañocazo is a minor Spanish variety for which the area is not recorded separately. Under recent planting regulations it is not listed for use in any wines of controlled appellation and not authorised for planting in any part of Spain. Scattered small plantings still exist in the sherry area and it is one of four varieties used as parents in a breeding program at the national research station at Jerez; the others are Palomino, Pedro Ximenez and an authorised variety called Garrido. It was probably brought to Australia in early collections from the sherry area and was formerly grown under the names of Palomino and Pedro. These names are used rather as family names in Spain, Garrido, for example, being sometimes called Palomino Garrio and the main sherry variety distinguished as Palomino Fino. The names in Australia were modified to Common Palomino and False Pedro or, more colloquially, Hardskin Pedro to avoid confusion with Palomino Fino or Pedro Ximenez. Cañocazo may also have been introduced from South Africa but the variety known there as False Pedro, of which there are several thousand hectares, appears to be another variety known as Rutherglen Pedro in Australia. This has now been identified as Dourado, which is probably the Pedro Luis of Spain. There are about 290 ha of Cañocazo in Australia, mostly in South Australia with a little in the Murray Valley Irrigation Areas of New South Wales and Victoria.

Cañocazo is a productive and fairly vigorous variety. The leaves are similar in shape to those of Pedro Ximenez but less deeply divided, being usually only moderately 3-lobed, somewhat darker green with bristles and occasional tufts of hair on the lower surface. The bunches are large, conical and loose without being straggly. The berries are medium, round to short oval, with a firm flesh and tough skin, golden in colour and neutral in flavour, somewhat reminiscent of Doradillo but ripening much earlier.

From its origin Cañocazo would be expected to be suitable for fortified wines and for distillation. It is more reliable viticulturally than Pedro Ximenez and perhaps also Palomino but does not have the same reputation for the quality of its wines.

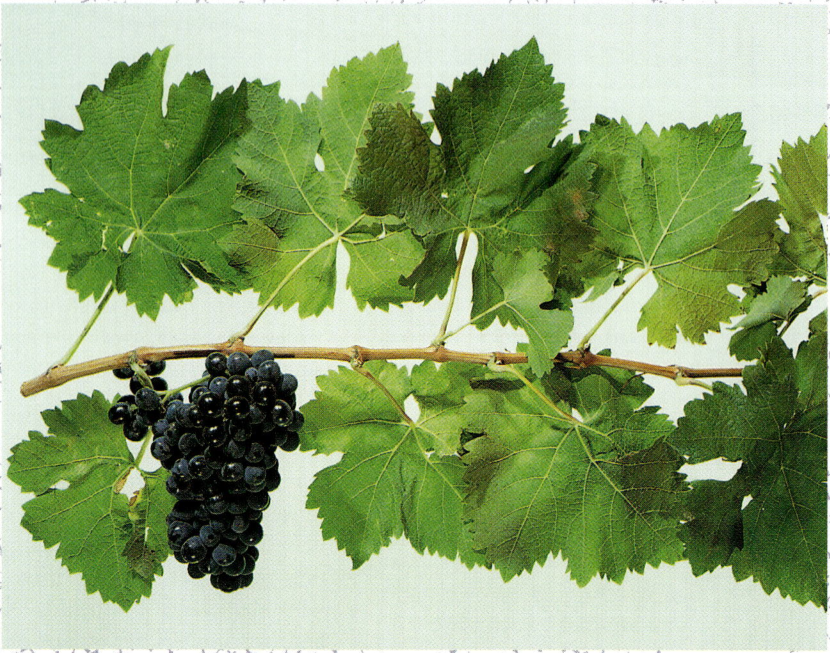
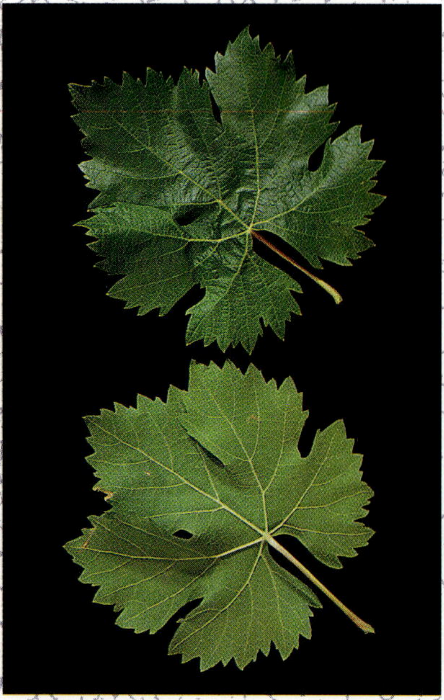
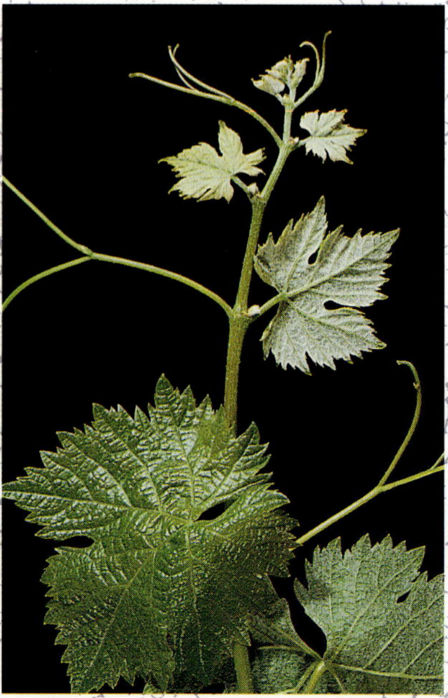

Carignan

Carignan is best known as a French variety, although it is of Spanish origin, taking its name from the town of Carinena in Aragon. The area planted in France increased from 170 000 ha in 1958 to 211 250 ha in 1968 but had decreased to 167 100 ha by 1988. It is a prominent variety in several areas in Spain but has not made much headway in other European countries. The former large area of Carignan in Algeria has been considerably reduced since 1962. It is still an important variety in California where it is known as Carignane, although the area has declined from 12 500 ha in 1974 to 3873 ha in 1992. It is grown to a lesser extent in Chile and Argentina. It has not been grown commercially in Australia and should not be confused with Bonvedro, which was formerly incorrectly called Carignan in South Australia.

Carignan is a vigorous and upright grower with large, thick 5-lobed leaves with tufted hairs on the lower surface. A distinctive feature is a pronounced puckering of the leaf blade near the point of attachment to the petiole, although this is also shown to some extent by another large-leafed variety, Tempranillo. Carignan has rather large, compact bunches with medium, short oval berries which are not readily damaged by rain during the ripening period. The firmly attached bunches, while easily harvested by hand, require considerable force in mechanical harvesting. The variety is more susceptible to fungal diseases than most other wine grape varieties and needs a higher level of light for full photosynthetic efficiency. In Australia, therefore, it should do best in the inland irrigation areas.

Carignan gives wines of moderate colour, good tannin and no pronounced varietal character. In France and Spain it is often crushed with varieties such as Grenache, Cinsaut and Tempranillo to give very pleasant wines ready for drinking when young.

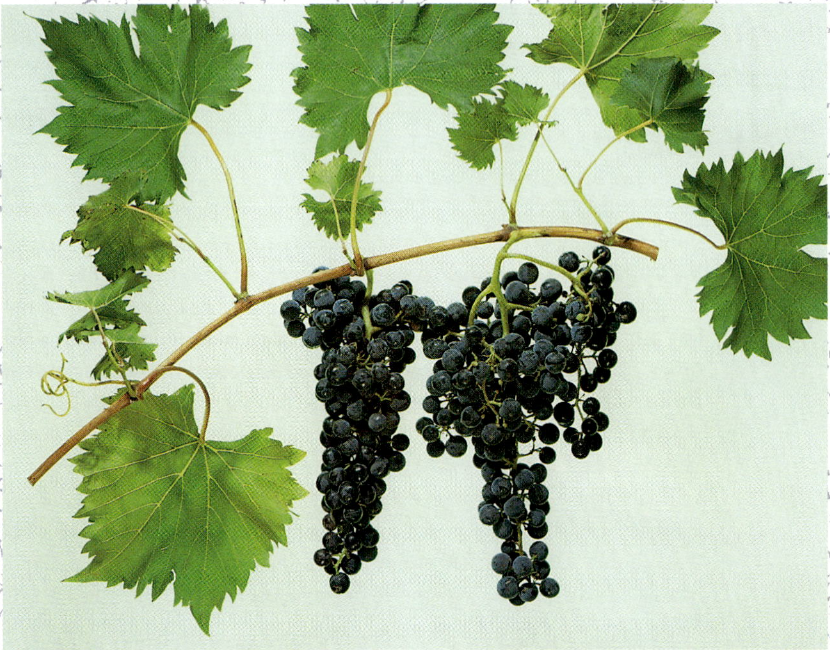

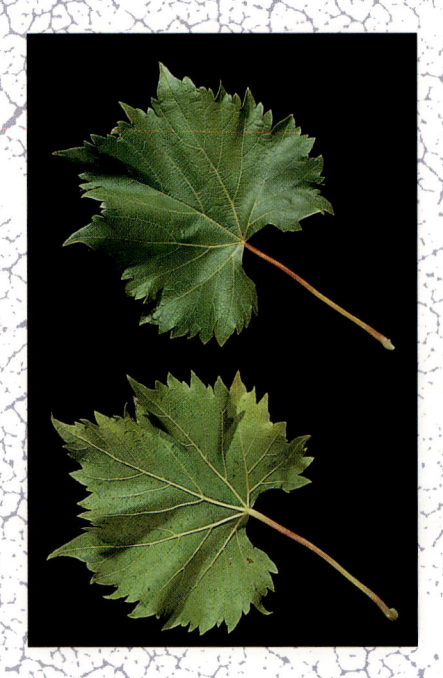

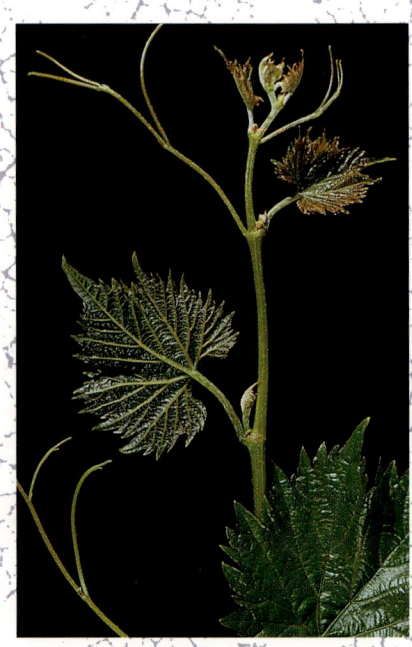

Chambourcin

Chambourcin is a complex hybrid produced in France by the private breeder Joannes Seyve (JS26-205). Its parentage has not been published but it would be based on the better Seibel hybrids and involve up to 8 of the American species of Vitis. It was officially released in 1963 and the area in France increased from 64 ha in 1958 to 3369 ha in 1979, but had decreased again to 1204 ha by 1988. It is being grown commercially in the eastern United States and was introduced into Australia by CSIRO in 1973.

Chambourcin is a vigorous variety with a spreading habit of growth. With often 3 and occasionally 4 bunches per shoot the fruit is noticeably well spread out on the vine. Leaves are medium with the petiole long in relation to the size of the leaf blade, mostly entire with coarse teeth, glossy above and with a few fine bristly hairs below, and with a very open petiolar sinus. Bunches are medium, well filled and cylindrical, usually winged, with medium, round berries with a heavy bloom. Chambourcin shows very good resistance to downy and powdery mildews, and to phylloxera, both against damage to the root system and against galling on the leaves

Wines from Chambourcin have drawn some favourable comments even in France, where the planting of hybrids is now heavily discouraged. Experimental wines in Australia have shown good colour and acidity and have scored well in blind tastings.

There were 36 ha of Chambourcin planted in Australia in 1990 of which 20 ha were not yet bearing.

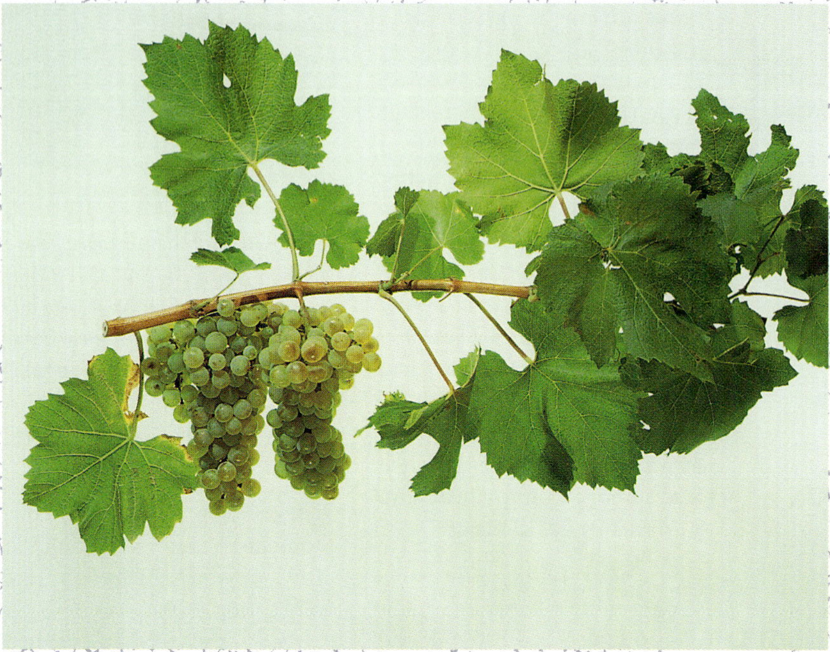
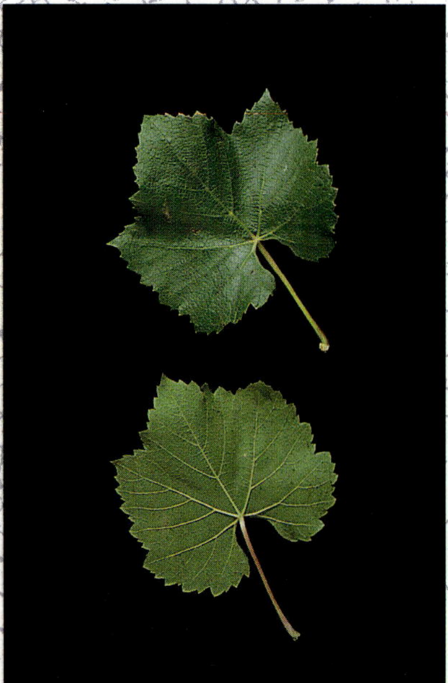
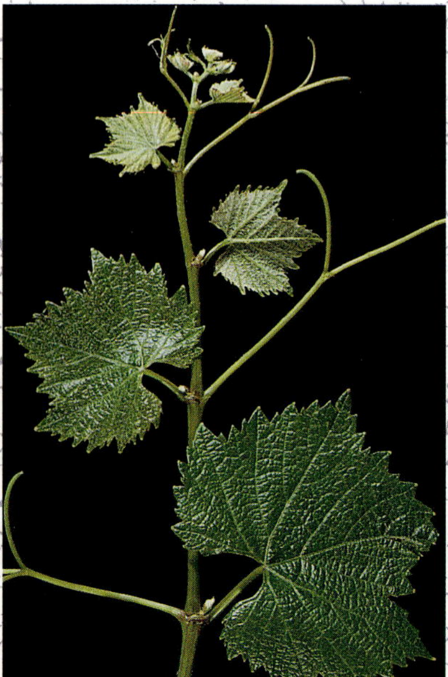

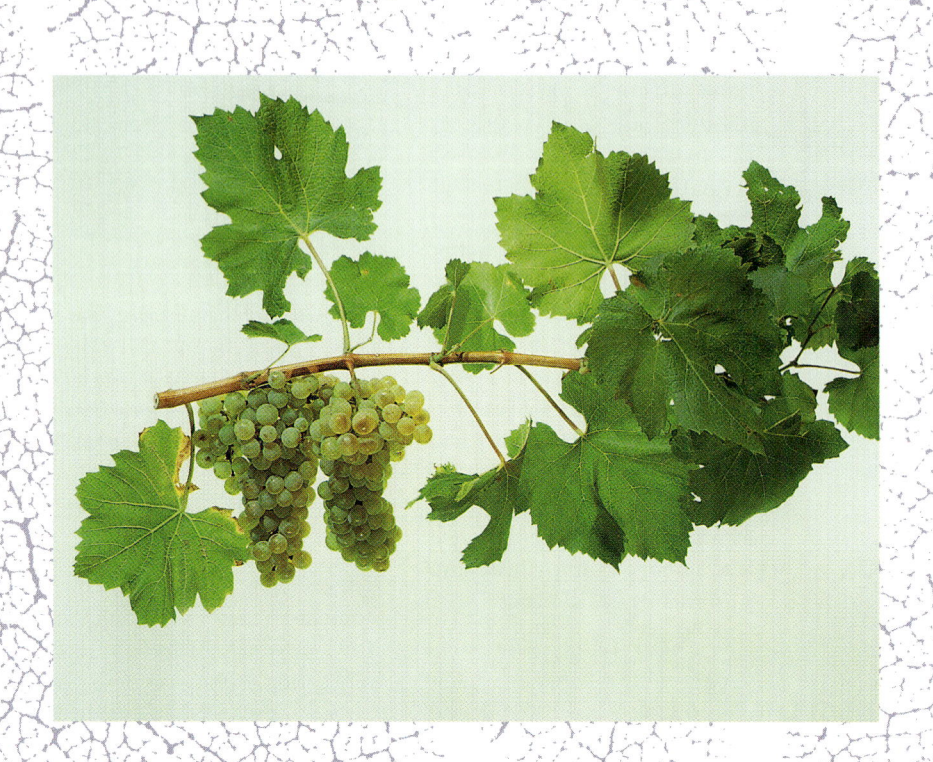

Chardonnay

To avoid confusion with Pinot Blanc, Chardonnay, not Pinot Chardonnay, is now the official name for this variety in France and California, two countries in which it is widely grown. The area in France increased from 7325 ha to 19 870 ha between 1958 and 1988, most of the plantings being in the Burgundy and Champagne regions. In California it was not widely grown until selected clones of high yield became available. Expansion since then has been rapid, with the area reaching 24 280 ha by 1992. Chardonnay is grown in many other countries but it is sometimes difficult to know how much confusion there may be with Pinot Blanc. There have been small plantings of Chardonnay in Australia for many years, but it is only recently that the variety has become popular. Of the 6107 ha planted by 1993, 1123 were still to come into bearing.

Clones of Chardonnay in Australia vary from moderately to quite vigorous. The leaves are medium in size, thick, undulating and rolling back a little at the edges, usually only slightly 3-lobed, and practically free of hair on the lower surface. The petiolar sinus is lyre-shaped and very characteristically cut right into the veins at the base.

Although the leaf is much the same shape as that of Pinot Blanc, it can be clearly distinguished by its other characteristics. The bunches of Chardonnay are rather small, cylindrical generally with a wing, well filled to compact, with small, round berries.

Chardonnay is used in the fine white wines of Burgundy and Chablis, and is one of the varieties used in Champagne. In California it is recommended for the cooler areas. The most suitable areas for the variety in Australia have still to be established but it appears to be possible to make high quality wines from grapes grown in warmer areas.

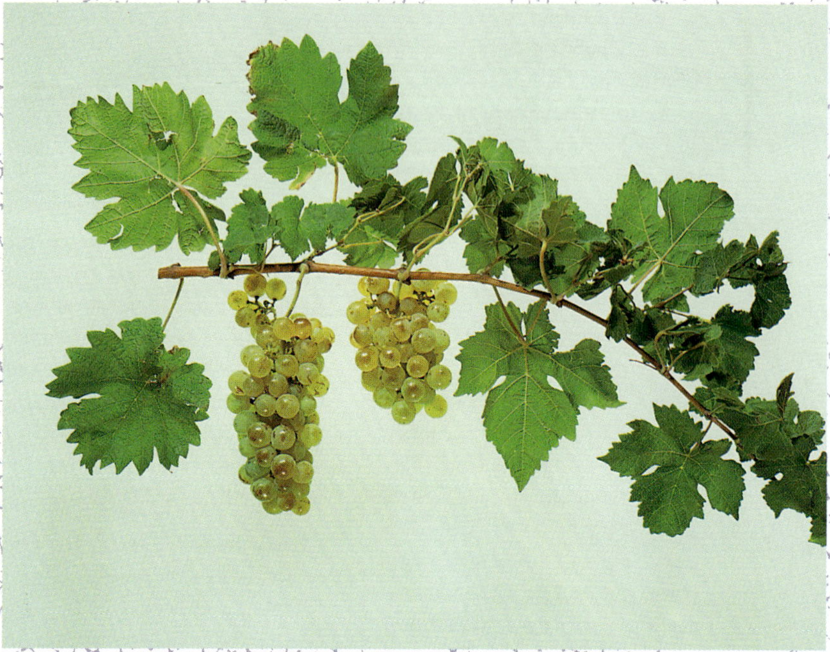

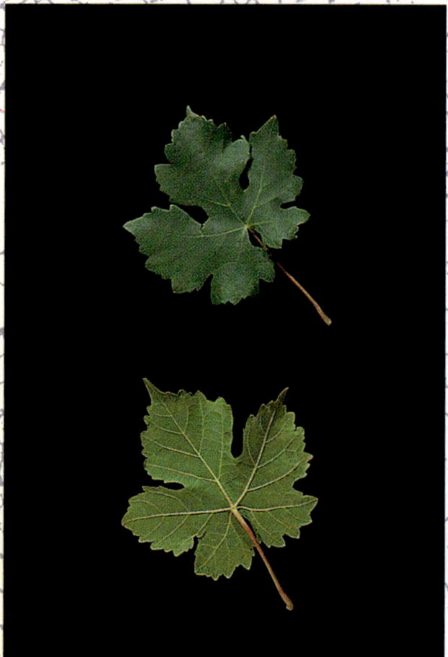

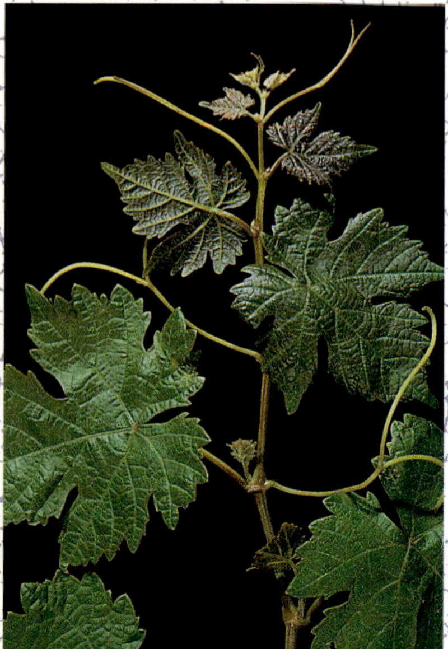

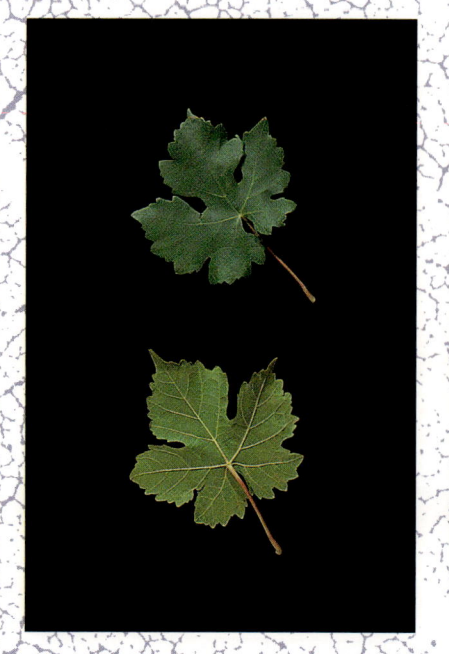

Chasselas

Chasselas is an important table grape in Europe. There are about 15 000 ha planted for this purpose in France and widespread plantings in Italy. It ripens early but it is popular in Europe even when other varieties are available. The pleasant flesh texture outweighs the presence of seeds in the berries which are rather small for a table grape. It is used as a wine grape in cool areas and may have originated in Switzerland where it is the principal white wine grape. There are about 800 ha in Alsace, 1400 ha in Germany and plantings also in Austria and eastern Europe. About 100 ha are recorded in Australia. In Victoria, which has about half the area, it is used for wine. In Western Australia and New South Wales, which share the rest, a fair proportion of the crop is marketed as table grapes. The variety is sometimes known as Golden Chasselas or Chasselas Dore, but Palomino has been mistakenly called Golden Chasselas in California and this mistake may have carried through into Australia.

A very characteristic feature of Chasselas is the bright, bronze-red colour of the shoot tips in spring. The mature leaves are clear-green, 5-lobed, and usually without hairs on the lower surface except along the main veins. In cooler areas it is reputed to be vigorous with large, fleshy, tangled tendrils. Bunches are medium in size and cylindrical with medium, round berries. At Merbein, the shoots are very short and stunted but the exposed fruit holds its condition well. Yields are only moderate and are made up of numerous rather small bunches. Because of the short shoots and high bud fruitfulness, the vines are pruned to very short spurs as for Muscat Gordo Blanco.

In cooler areas Chasselas is used to produce pleasant light table wines with no pronounced varietal character. In hot areas, the fruit is too low in both sugar and acid to be satisfactory for winemaking.

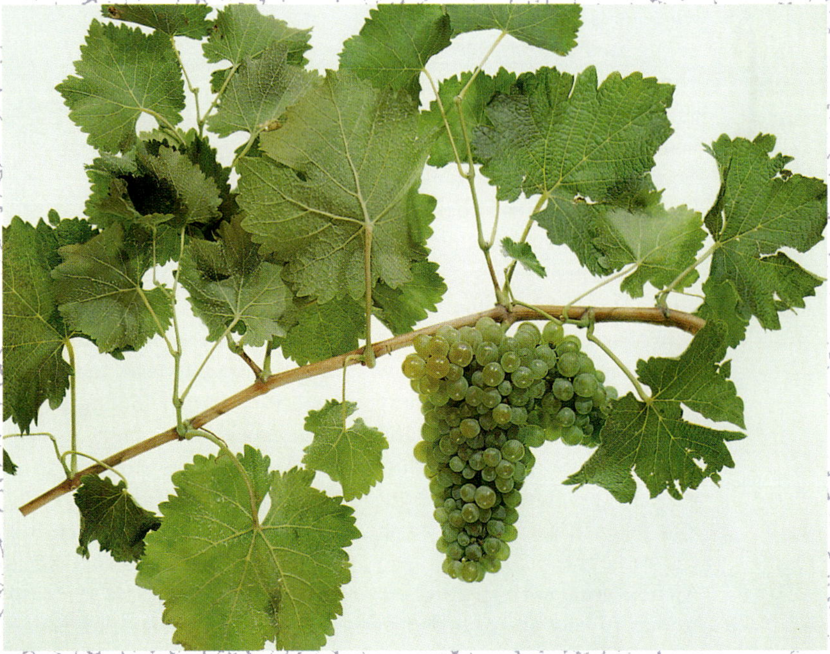

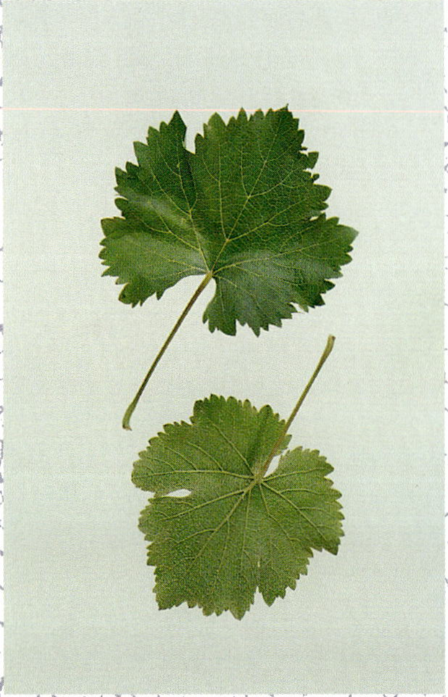

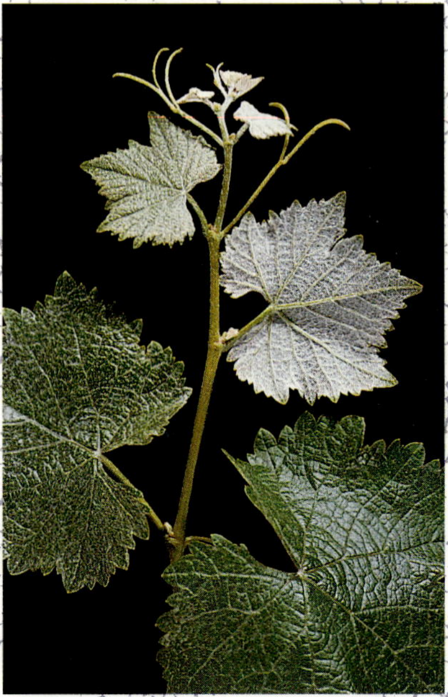

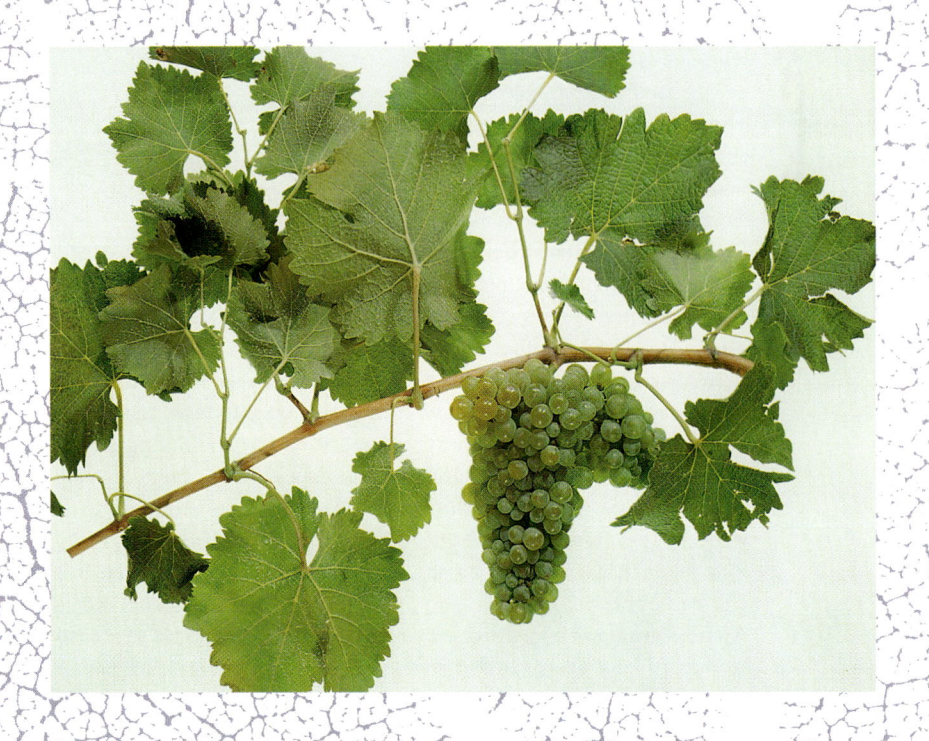

Chenin Blanc

Chenin is the official name for this variety in France, but it is often called Pineau de la Loire. It is the main variety of the Loire Valley where there has been a small decrease in the area planted, from 14 200 to 9050 ha, between 1968 and 1988. In California there has been a rapid increase to 11 700 ha in 1992. The variety wrongly called Pinot Blanco in Chile and Argentina is really Chenin Blanc and 4000 ha are recorded in Argentina. The largest area of Chenin Blanc is in South Africa where 27 700 ha are planted under the name of Steen. The variety seems to have been brought to Australia under several names and its identity lost. Of a total of 670 ha, there are 100 ha in Western Australia, where it was formerly incorrectly known as Semillon, and 430 ha in South Australia, where it was wrongly named Albillo or Sherry.

Chenin Blanc is a vigorous variety with an early bud burst. One reason for its popularity in South Africa is that its shoots adhere firmly and are not easily blown off by high winds in spring. The leaves are olive green, rough, undulating, 3- to 5-lobed with tufted hairs on the lower surface. Bunches are medium, conical, winged and compact with small, oval berries. It shows clonal variation in bunch shape and time of maturity and at least some clones are susceptible to splitting by rain and attack by bunch rot at harvest

Chenin Blanc has a good acidity and is used very successfully in South Africa and California for well-balanced dry table wines. In favoured parts of the Loire Valley it can be affected by 'noble rot' and produces excellent luscious sweet wines. It is also used in some very good sparkling wines.

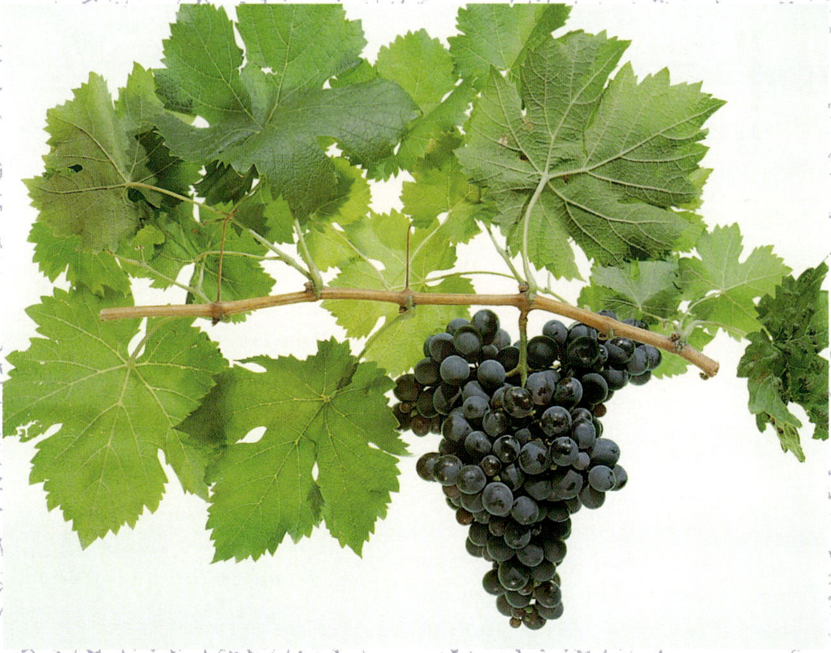
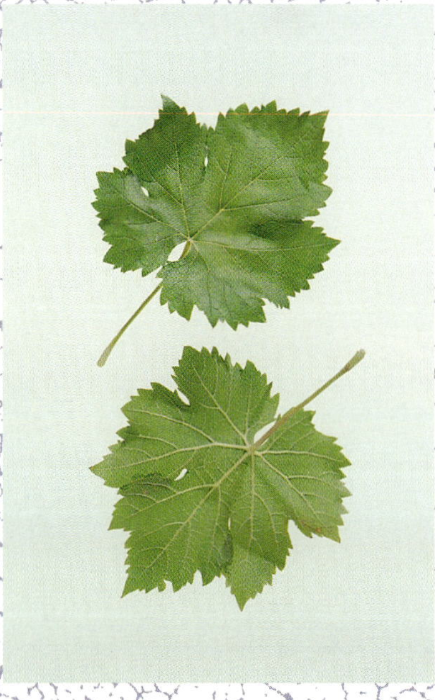
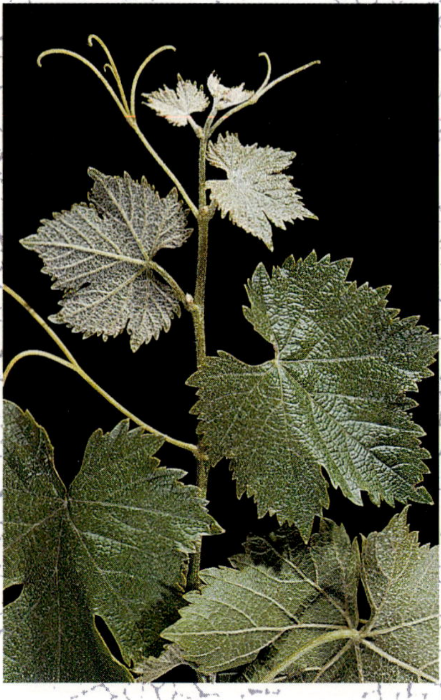

Cinsaut

Cinsaut, sometimes spelt Cinqsaou or Cinsault, is a variety from the Mediterranean region in the south of France. It has become confused with another variety Oeillade (also spelt Ouillade, Ulliade, Ouillard) which, because it sets less reliably than Cinsaut, has practically disappeared from France since the phylloxera invasion. Oeillade was used as a table grape and to replace it Cinsaut has been called Oeillade when sold as a table grape. About 760 ha of Cinsaut are registered for the production of table grapes but its main importance is as a recommended variety for improving the quality of the wines from the south of France. Plantings for wine increased from 18 200 ha in 1968 to 48 200 ha in 1988. There are also about 4500 ha in South Africa, where it is called Hermitage. There were formerly about 60 000 ha in Algeria and, being the black variety best adapted to the hot dry conditions, it may not have been as drastically reduced in area as other varieties. It is also grown in Italy under the name of Ottavianello with about 3000 ha in the province of Brindisi. In California, where it is called Black Malvoisie, there are about 40 ha. In Australia there were 59 ha by 1990, a little over two-thirds of this being in South Australia and the rest evenly divided between Victoria and New South Wales. It has been called Blue Imperial in north-east Victoria, Black Prince at Great Western and often Ulliade or Oeillade in other areas.

Cinsaut is a fairly vigorous variety with a spreading habit of growth. It has medium, rather rough, distinctly 5-lobed leaves, with tufted hairs on the lower surface varying in amount from slight to moderate on different clones. The petiolar sinus is usually a narrow lyre. Cinsaut has medium to large, well filled to compact bunches with long stalks. The medium, oval berries have a tough skin and firm texture and are not easily split by rain when ripe. The berries detach fairly easily and the variety is very suitable for mechanical harvesting. Cinsaut is rather sensitive to fungal diseases, although less so than Carignan. It can be pruned to spurs or canes.

By itself Cinsaut gives wines which have an attractive red colour and agreeable bouquet but which are low in tannin. It is generally used as an element of smoothness in conjunction with other varieties such as Grenache and Carignan in France, and Primitivo in Italy. In France it is used in many wines of controlled appellation including Cotes du Rhone, Tavel, Chateauneuf-du-Pape, Cassis and Bandol.

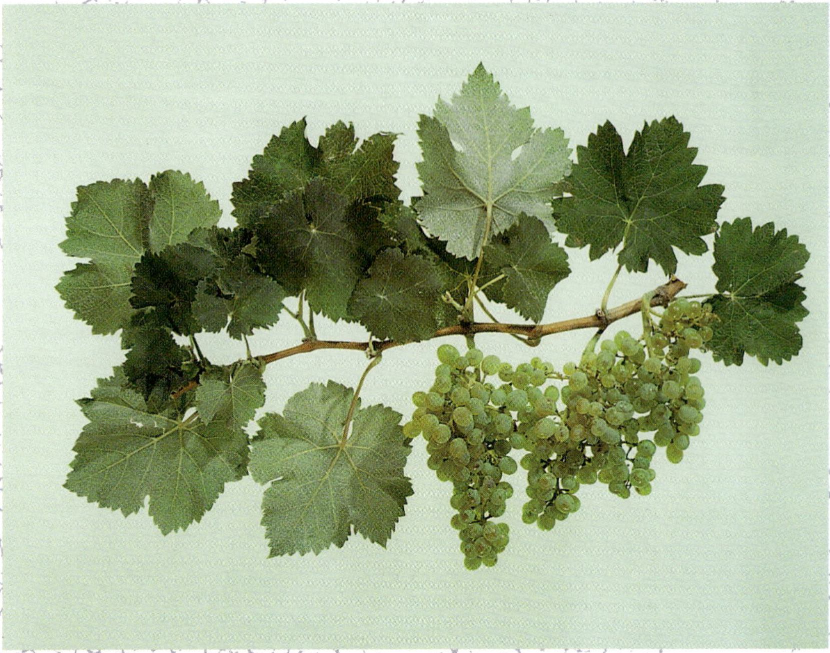

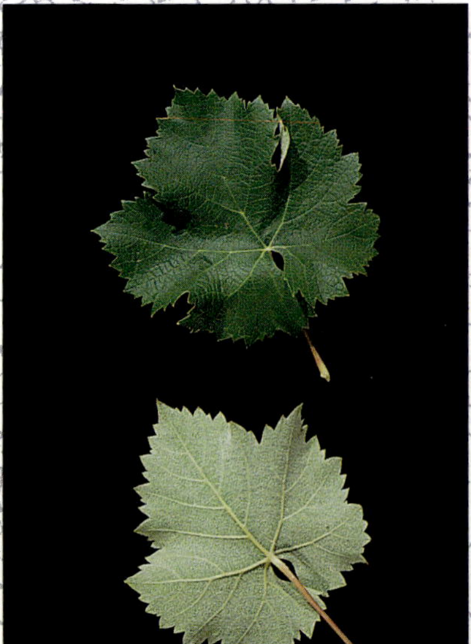

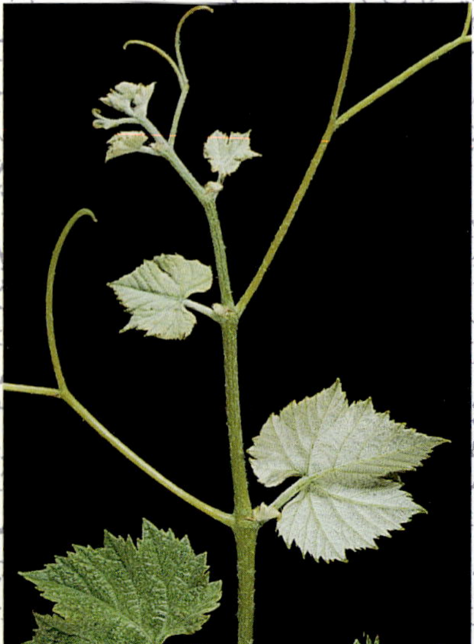

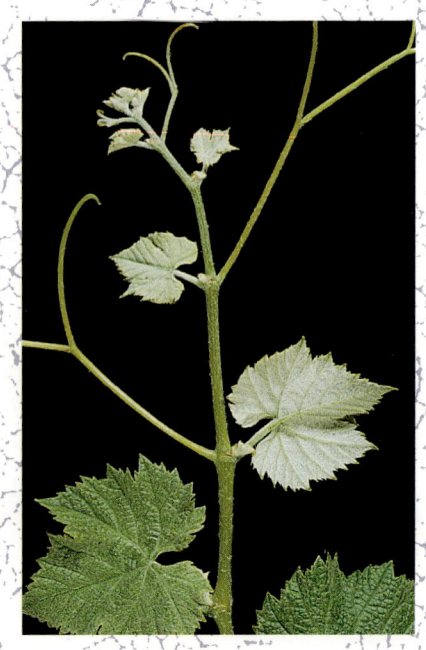

Clairette

Clairette is an important variety in the south of France. The area of plantings has decreased from 12 500 ha in 1968 to 3900 in 1988. It was the most important white variety in Algeria before independence with perhaps 10 000 ha. It does not appear to be grown in Europe outside France and the only major plantings elsewhere appear to be in South Africa, where there are about 2150 ha. In Australia there are about 26 ha, nearly all in the Hunter Valley of New South Wales. In France it is sometimes known as Blanquette and this name is used in New South Wales. It should not be confused with the so-called Blanquette of South Australia, which is a clone of Doradillo. Two other varieties in France have also been referred to as Clairettes; Ugni Blanc (syn. Trebbiano) is sometimes called Clairette Ronde in both France and Italy, and Bourboulenc has been called Grosse Clairette or Clairette Doree.

Clairette is a vigorous variety with the young shoots liable to be blown off by high winds in spring. It has medium, 3- to 5-lobed, rough and undulating leaves with the petiolar sinus usually closed, often with overlapping edges. The leaves are distinctly thick and leathery and downy white on the lower surface. Clairette ripens late and has medium, well filled bunches and medium, oval berries with rather thin skins.

Oxidation occurs very readily during winemaking with Clairette, and 'rancio' wines made with over-ripe grapes are still covered by the French controlled appellation regulations. Clairette also imparts a distinctive varietal character in the absence of oxidation and is used in many appellation wines, both still and sparkling. It is sometimes used alone and sometimes in combination with other varieties; some red wines contain up to 20% Clairette.

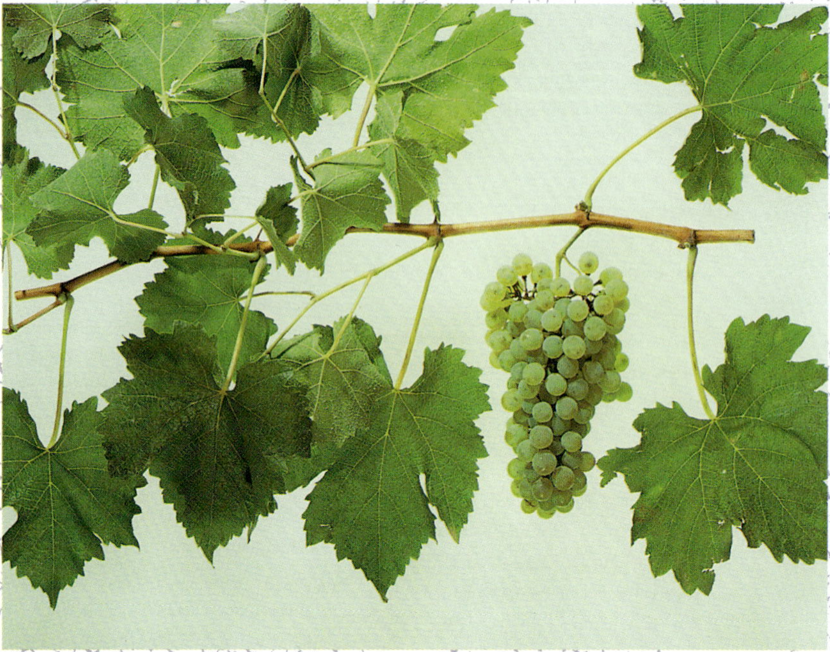
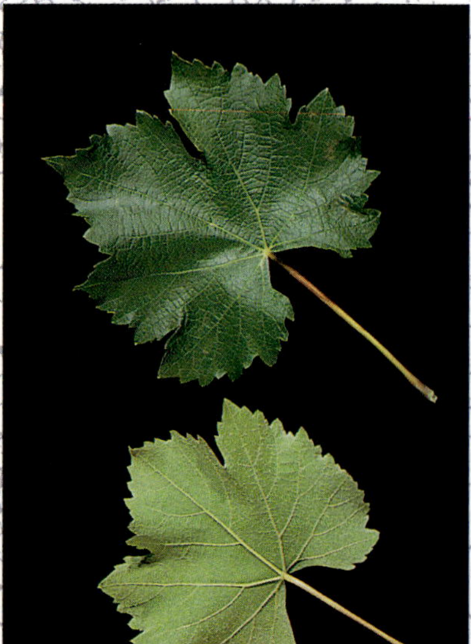
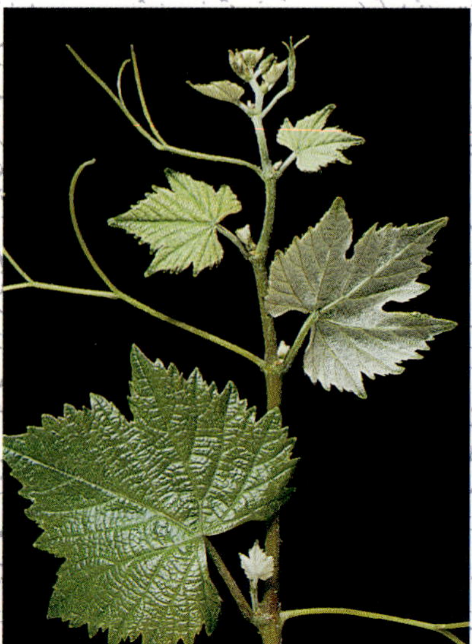

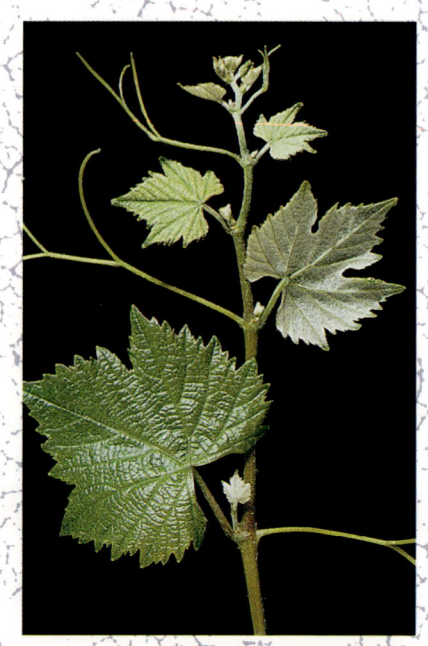

Colombard

Colombard is a variety from the Bordeaux region of France, grown mainly on the east of the Gironde. It appears to have lost favour in recent years, with plantings declining from 11 892 ha to 5000 ha between 1968 and 1988. On the other hand it has seen a remarkable expansion in California, with nearly 21 400 ha planted by 1982, followed by slower expansion to 21 890 ha by 1992. There are aboout 9000 ha in South Africa. A number of introductions into Australia from California have been made over the last thirty years and by 1983 there were about 360 ha planted This had increased to 887 ha by 1993.

Colombard is a very vigorous variety which without balanced management can produce thick solid canes rather brittle for cane pruning and inconvenient for spurring. The buds burst early and the young shoots are liable to be blown off by strong winds in spring. The leaves are medium in size, thick, only slightly lobed, tending to fold in from the edges and with tufted hairs on the lower surface. The petiolar sinus has a distinctive open V shape. The bunches are medium to fairly large, well filled, more or less cylindrical and sometimes winged, with rather short, thick stalks. Berries are medium and almost round. The fruit will remain on the vine in good condition after it is ripe and retain a good acidity, but if left too long may give an unpleasant character in its wine.

In France, Colombard is regarded as only an accessory variety for table wine. It will give a brandy of high quality, but not superior to that from Folle Blanche or Saint Emilion (Trebbiano). In California, Colombard, because of its acidity, it is considered suitable for blending to produce quality white table wines, and also for producing sparkling wines.

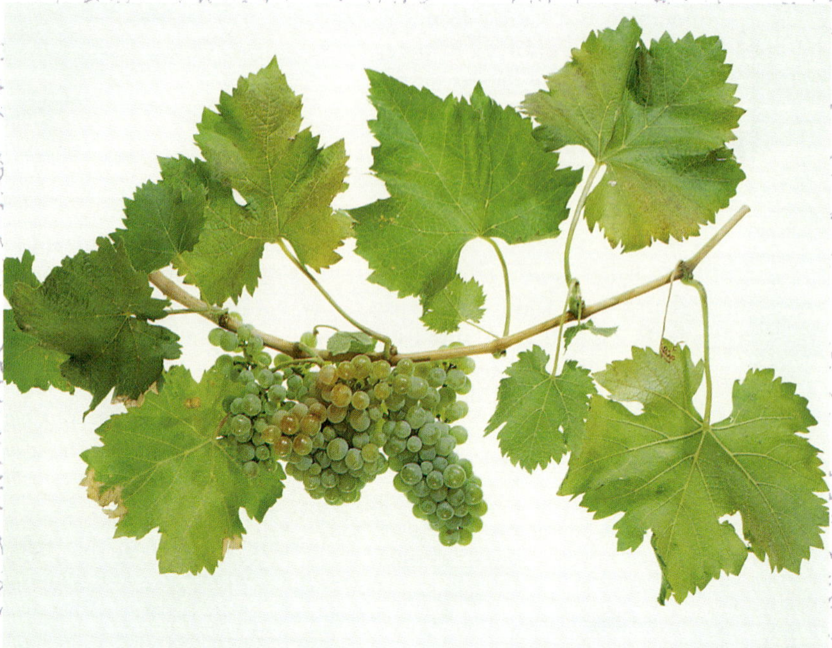

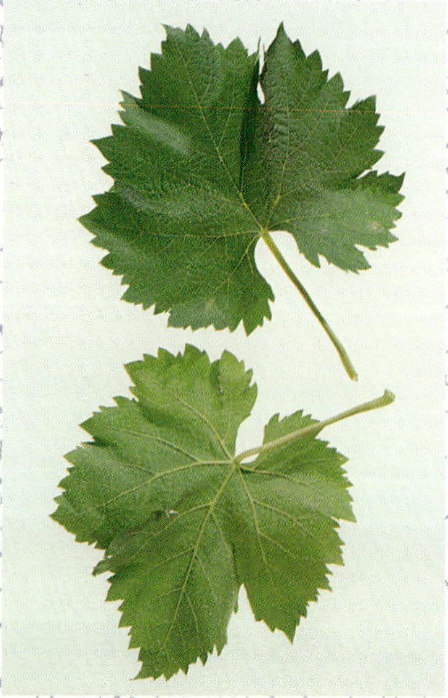

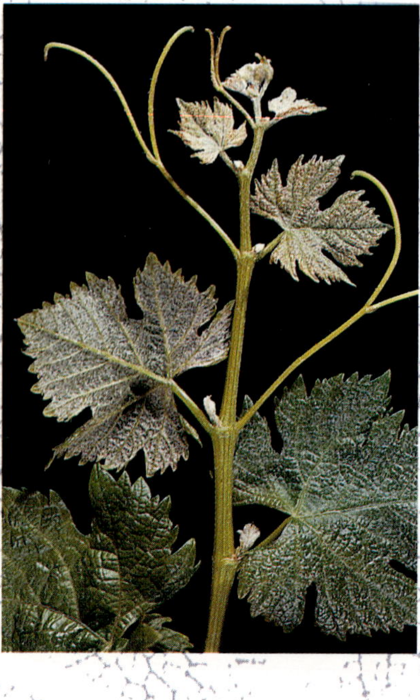

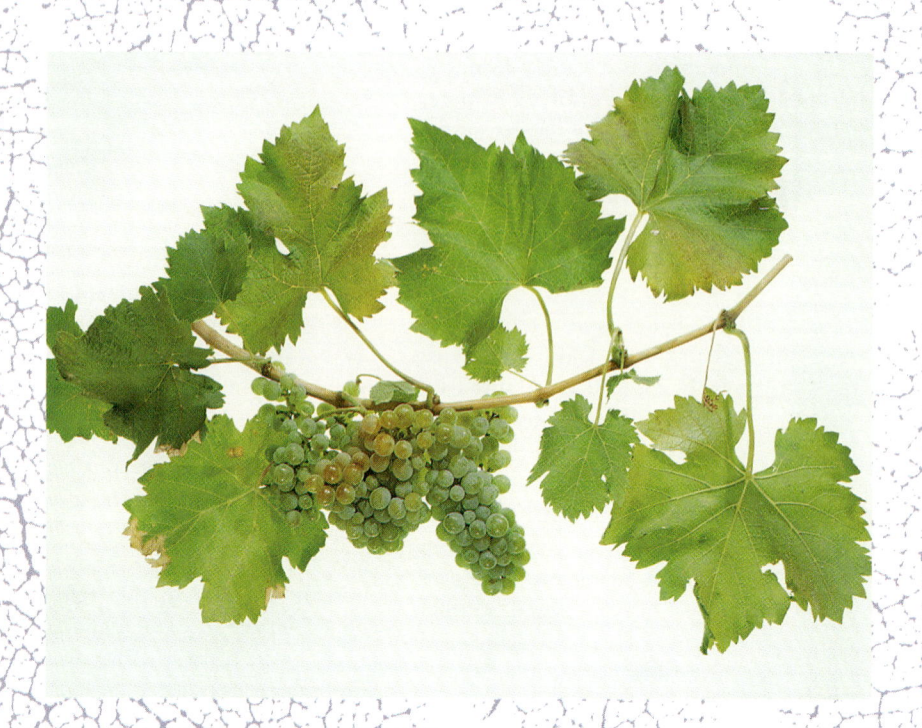

Crouchen

Crouchen is a French variety which has now practically disappeared from France. It is still a recommended variety in the south-west near the Pyrenees and appears on the list of varieties which can be used in the wines of Bearn. However with the use of rootstocks to combat phylloxera, Crouchen became more susceptible to foliar fungal diseases and bunch rot, and other varieties such as Baroque and Manseng Blanc, better adapted to the wet and humid climate of the area, have been preferred. Crouchen was also used in the excellent 'vin de sable' produced near Capbreton before the afforestation of the Landes displaced the vineyards last century. The name used there for the variety, Sable Blanc, may well correspond to the 'Sales Blanc' introduced into New South Wales by James Busby in 1832.

The variety now appears to be grown only in South Africa (about 3500 ha) and Australia (231 ha). In South Africa it somehow came to be called Riesling and forms most of the plantings under that name in the country. It seems to have been introduced from South Africa to the Adelaide area as Riesling and to have spread from there to the Clare and Riverland areas before the mistaken identity was discovered. It was then called Clare Riesling until finally identified as Crouchen. It was also brought into the Barossa Valley, where it was mis-identified as Semillon, and taken from there to the Sunraysia area as Semillon before it was identified as Clare Riesling.

Under favourable conditions Crouchen is a vigorous variety with a trailing habit of growth. It is one of the most sensitive varieties to attack by root knot nematode. The leaves are dark green, rough and undulating, 5-lobed and tending to roll back at the margins. The bunches are small to medium, usually winged, and compact. The berries are rather small, round to slightly oval, retaining a definite green colour when ripe, with a coppery tint where exposed to the sun.

Crouchen does not seem to adapt successfully to as wide a range of climates as Rhine Riesling. It does not ripen in a very cool area such as at Geisenheim in the Rhine Valley and loses its character in a hot climate. In areas to which it is adapted, it produces pleasant dry white wines with a delicate varietal character.

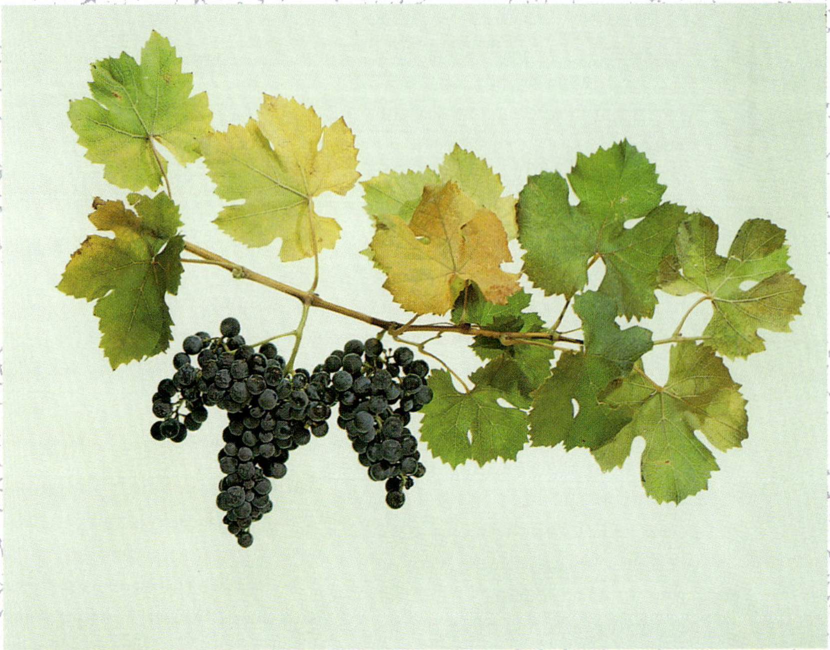
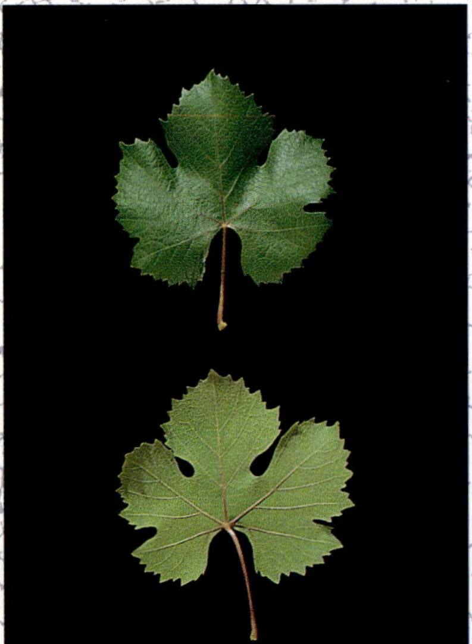
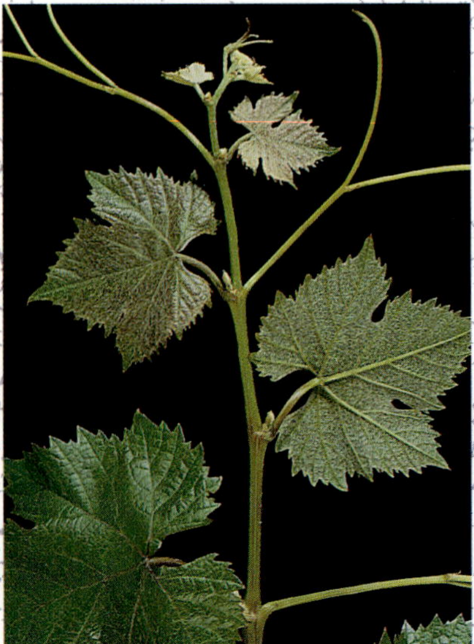

Dolcetto

Dolcetto is an important variety in the Piemonte region of Italy, the area planted possibly approaching 10 000 ha. Unlike some of the other red varieties of the region, which are grown also in other parts of Italy, Dolcetto is regarded as having a special adaptation limited to Piemonte. Despite this, it ranks about eighth in production among the red wine grapes of Italy. There appears to be very little Dolcetto in other countries, with only a few hectares in both Argentina and Australia. Australian plantings are confined to South Australia and Victoria. The area in South Australia, where the variety has been correctly named, has decreased over the years, while in Victoria, where it has been confused with Malbec, there have been a few small new plantings which were intended to be Malbec.

In Italy, Dolcetto is used alone in the denomination of origin wines Dolcetto d'Acqui and Dolcetto d'Ovada, which are well regarded wines of medium body and characteristic flavour. Wine from Dolcetto has a fine, bright red colour, which may, however, lack intensity if the vines are heavily cropped or grown in unsuitable conditions. There appear to be situations in the cooler areas of Victoria which are well suited to the variety.

DNA Dolcetto is a rather weakly growing variety with small leaves which take on a most attractive pink autumn colouring. The leaves are 3- to 5-lobed with an open V-shaped petiolar sinus and tufted hairs on the lower surface. Bunches are small to medium, shouldered and well filled with small, round berries which have a heavy bloom.

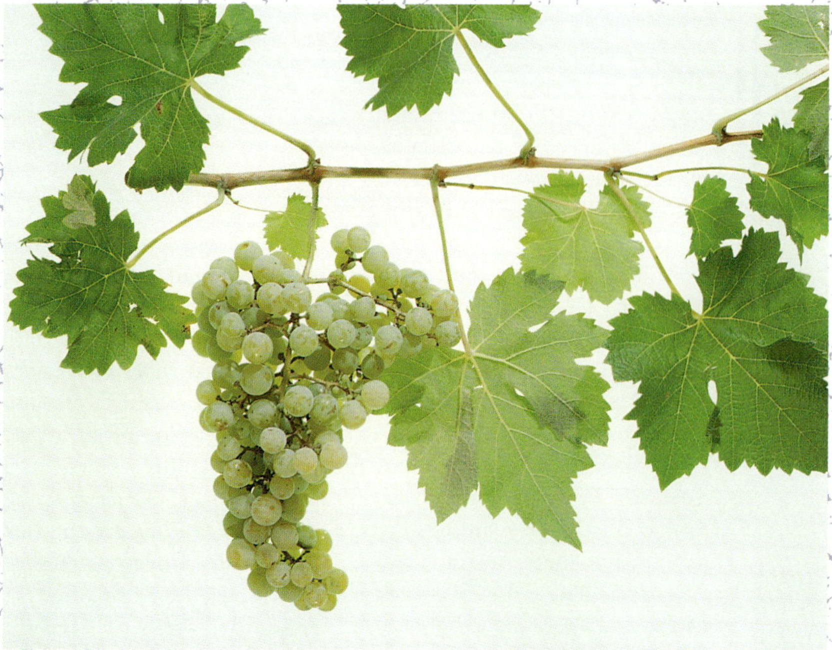

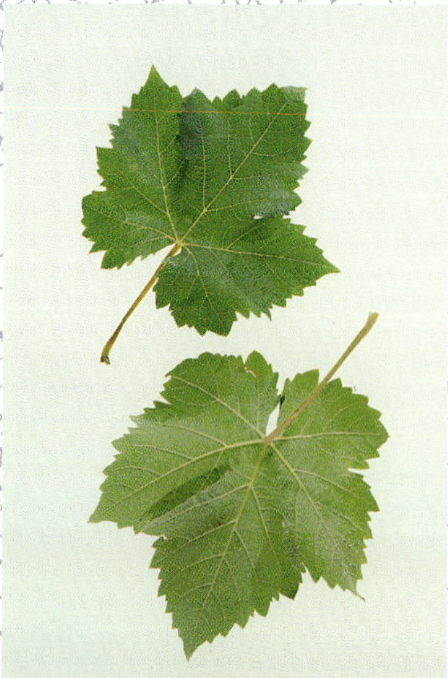

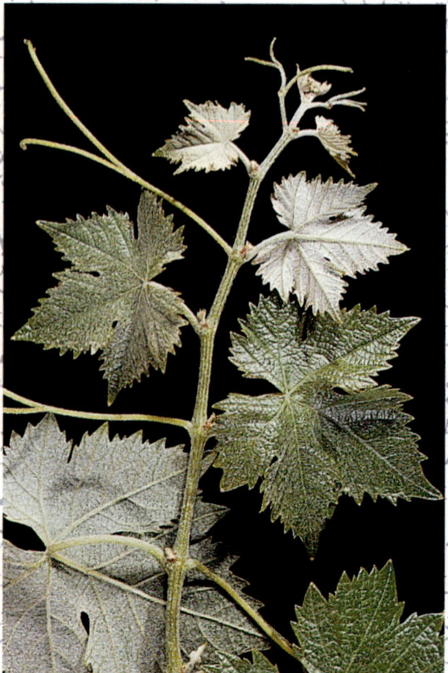

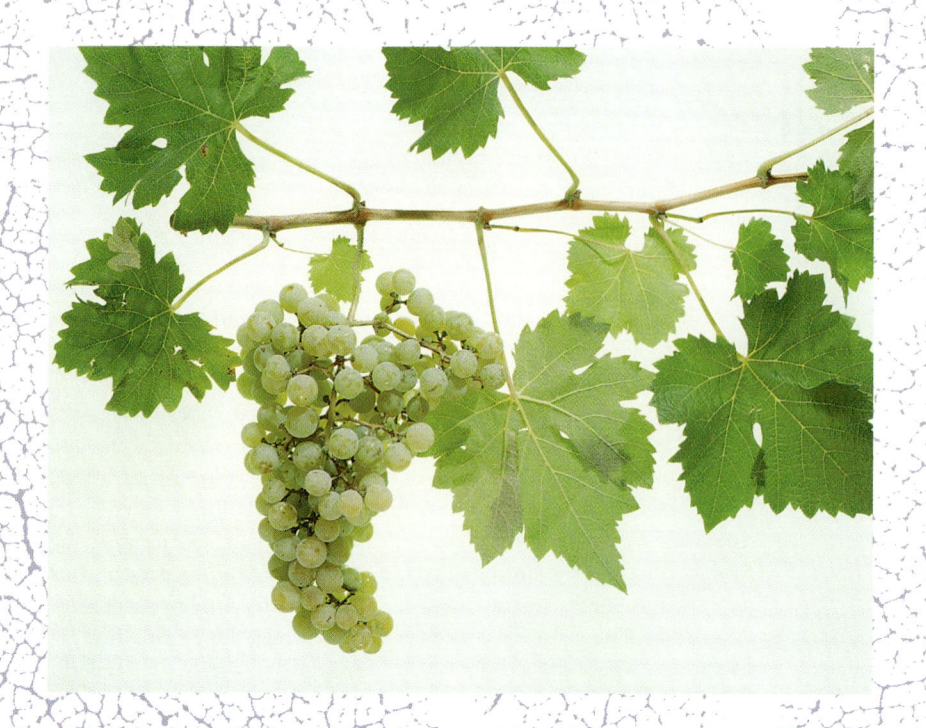

Doradillo

Doradillo is a variety which seems to have found its greatest acceptance in Australia. It was imported under that name from Spain by James Busby in 1832. It also appears to have been taken to South Africa and further importations were made to Australia from there. It does not seem to be a commercial variety in South Africa and it is not recorded separately in Spanish plantings. It may perhaps be included under Jaen, a name said to have been used for more than one variety in Spain. Jaen imported to Merbein from the Estacion de Viticultura y Enologia at Requena in Valencia province has proved similar to, but distinct from, Doradillo. There are about 645 ha of Doradillo in Australia, about three-quarters of it in South Australia and the rest divided between New South Wales and Victoria.

DNA Doradillo is a moderately vigorous variety, but has a rather shallow and weak root system which seems to be more sensitive to poor drainage than most other varieties. It has leaves which are dark green and rough above and hairy below, mostly 5-lobed with the lobes overlapping and short rounded teeth. The bunches are large and well filled. The berries are medium, round and firm, have a heavy bloom and become golden where exposed to the sun. The variety yields consistently well but the fruit matures late and is quite neutral in flavour.

In Australia, Doradillo is used mainly for distillation and for the production of sherries.

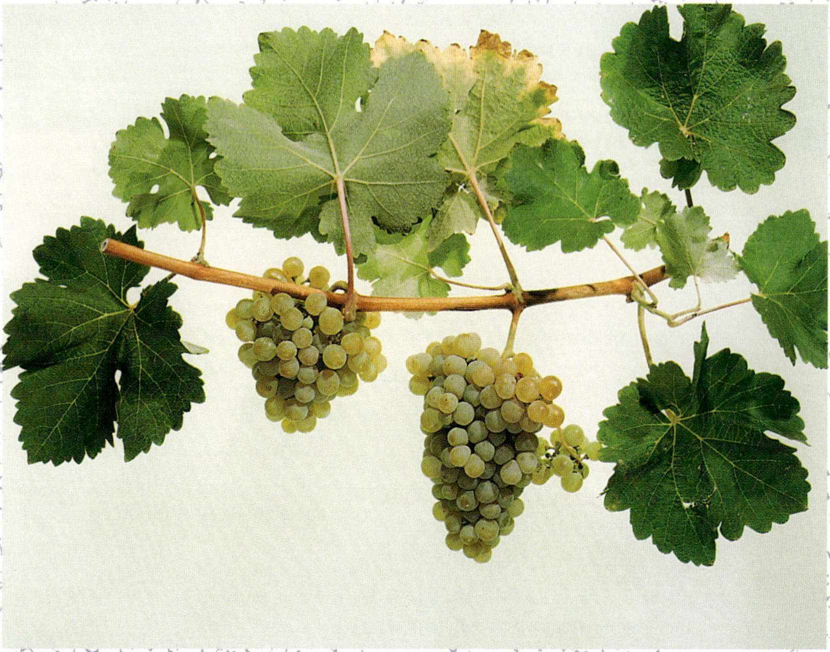

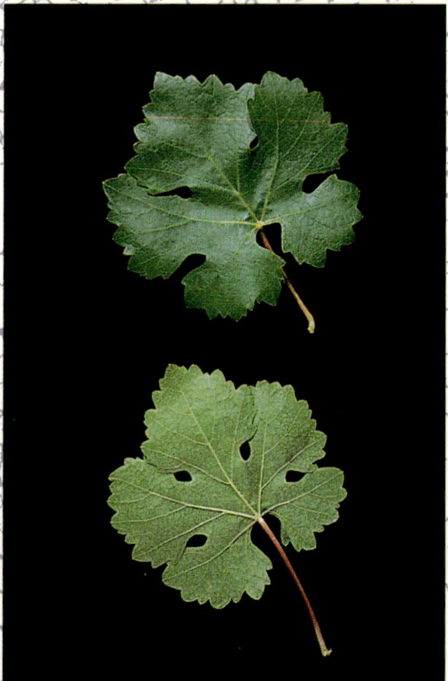

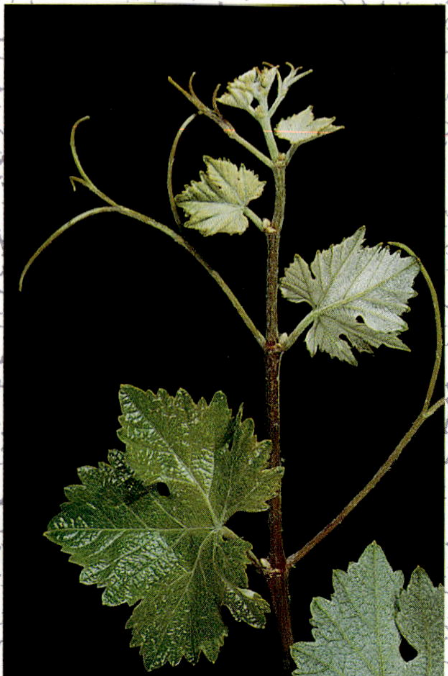

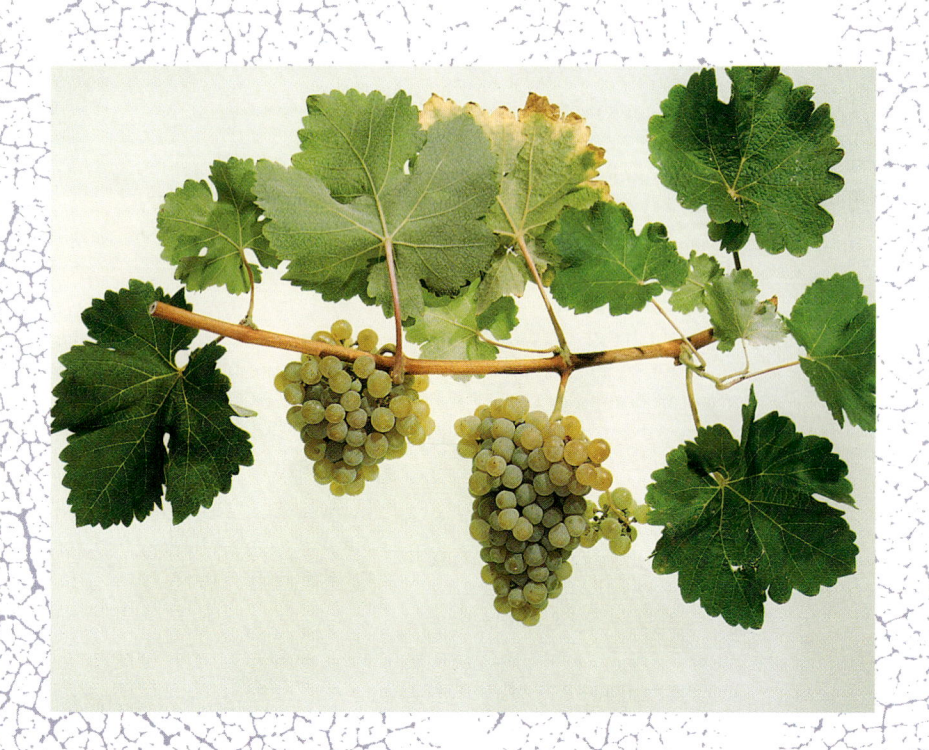

Dourado

Dourado is a variety from the Tagus Valley of Portugal. Perhaps its full name Gallego Dourado should be used because the name Dourado is sometimes also used for a different Portuguese variety Loureiro. In South Africa, about 3 500 ha are planted as (False) Pedro. It is likely that it was brought from South Africa to Australia as it was known here as Pedro, sometimes Pedro Ximenez in error, but more usually distinguished as Rutherglen Pedro. It may also have come in James Busby's collection in 1832 under some other name or unnamed as it occurred unidentified in vineyards which might be traced to this source. Dourado can be found as odd vines in most of the older wine growing districts. Appreciable numbers of vines are mixed with other varieties at Mudgee and Great Western and small plantings are present at Roma, and in the Murrumbidgee Irrigation Area and north-east Victoria

Dourado is a vigorous variety of upright growth habit. Although some leaves are almost entire, Dourado can be recognised by the presence of very characteristic leaves with five lobes which tend to overlap at their tips, leaving the sinuses as a series of five holes. All leaves have very rounded teeth, a few cobwebby hairs on the upper surface and more densely felted hairs on the lower surface. Bunches are small to medium, conical, well filled, often with a woody stalk, and have medium, round berries with a heavy bloom.

Because of the confusion with Pedro, the variety has probably been mostly used in fortified wines in Australia but it can be successfully used for dry white wines.

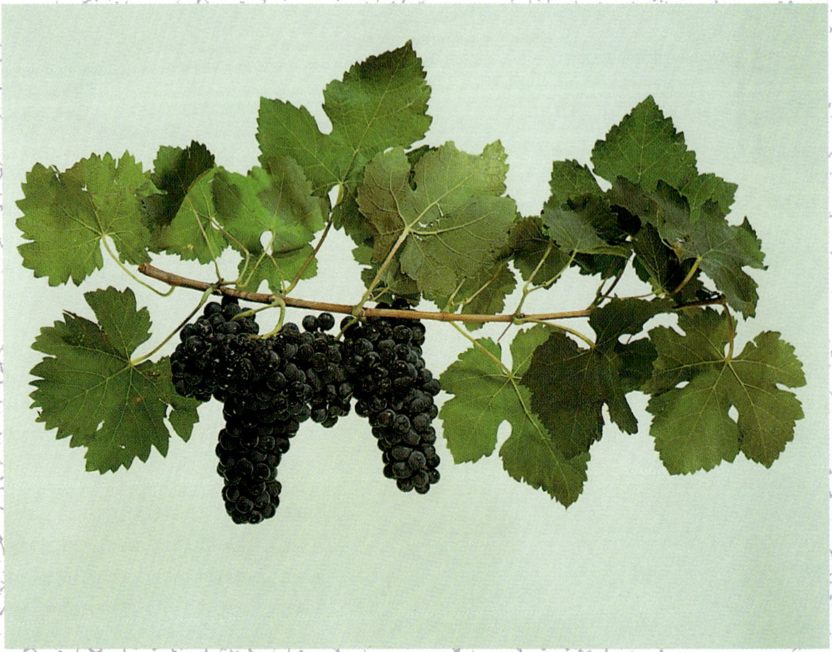

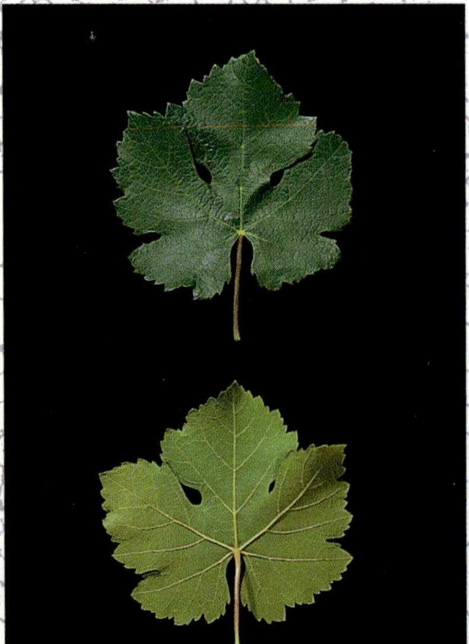

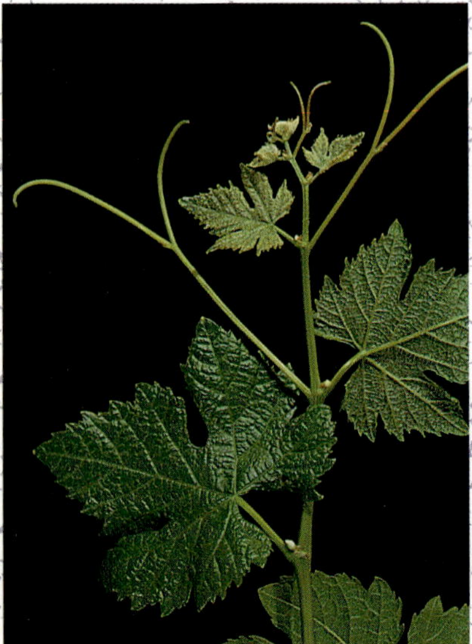

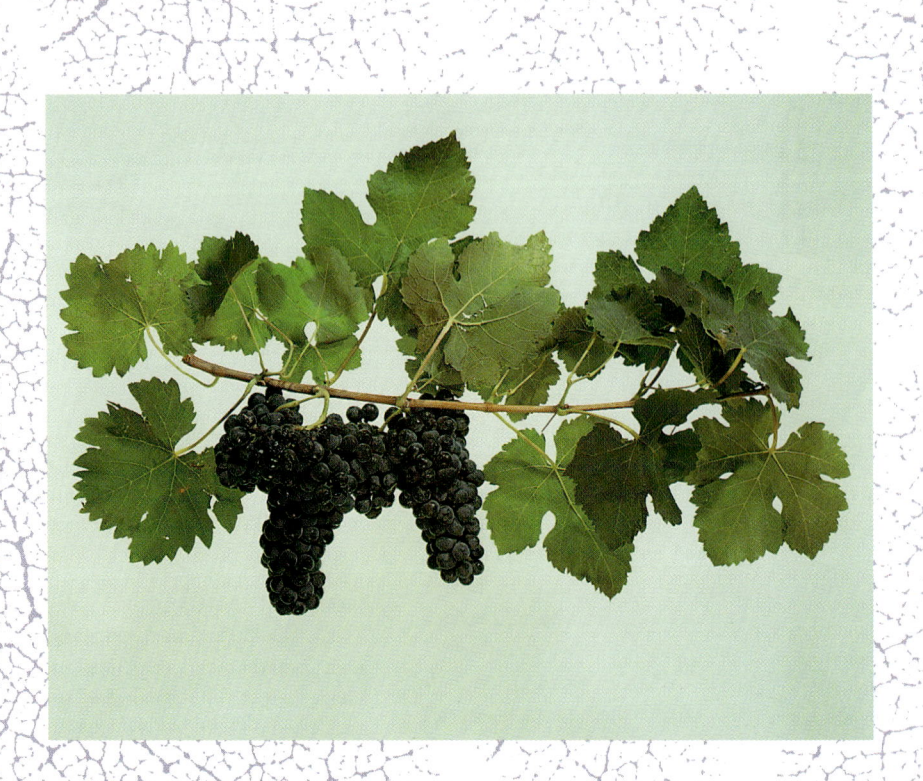

Durif

Durif has been known as a variety for only about a century. It owes its name to a Dr Durif who was propagating it in the Rhone Valley in France around 1880. It resembles Peloursin and may be a seedling or a sport from this variety. It proved popular as it appeared to have some resistance to downy mildew. However, it is no longer recommended or authorised and the area in France is declining. It is sometimes known as Pinot de l'Ermitage but is in no way related to the true Pinots. Another name, Syrah Forchue, refers to its tendency to produce forked shoots. The Petite Sirah of California, of which 1070 ha are recorded, is a mixture of Durif and Peloursin, in what proportions it is not clear. All clones imported to Australia have proved to be Durif. Australian crushings in 1990 included 108 tonnes reported as Durif but about half of this may have been Peloursin. Vineyards examined have shown roughly equal numbers of the two varieties.

Durif is a vigorous variety with a spreading habit of growth. It has medium to large, 5-lobed leaves, hairless above and with, at most, only a few cobwebby hairs below. Bunches are medium and compact, berries small and round when not pressed out of shape and with little bloom. Exposed fruit is sensitive to sunburn.

Durif produces wines of intense colour and high tannin requiring long aging. Australian wines would be less tannic because of the presence of Peloursin and this would need to be borne in mind if replacement of plantings was contemplated.

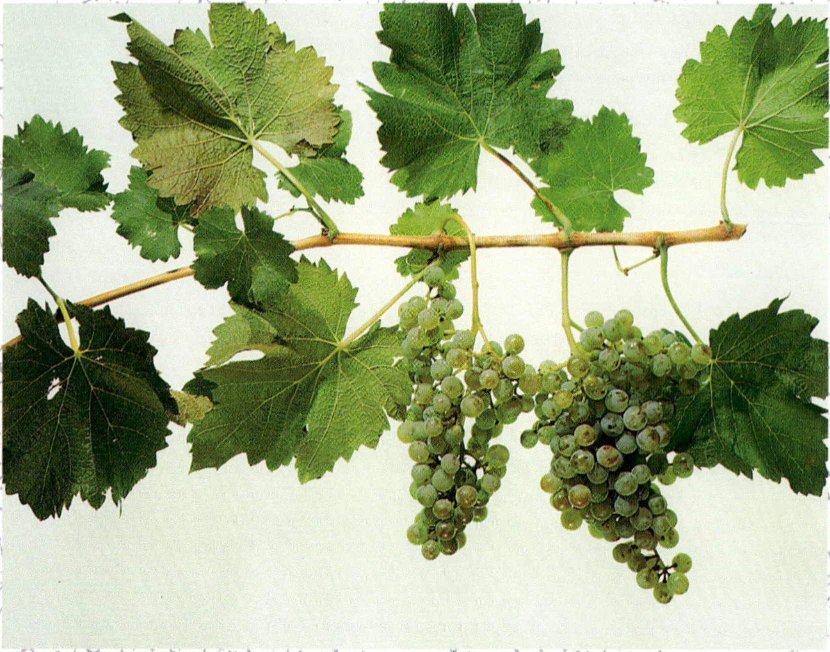

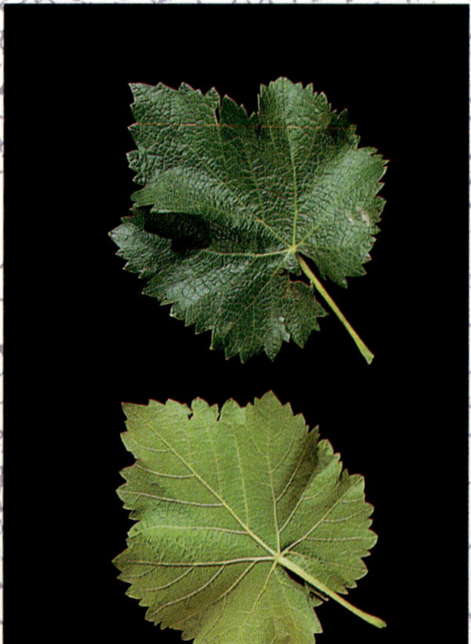

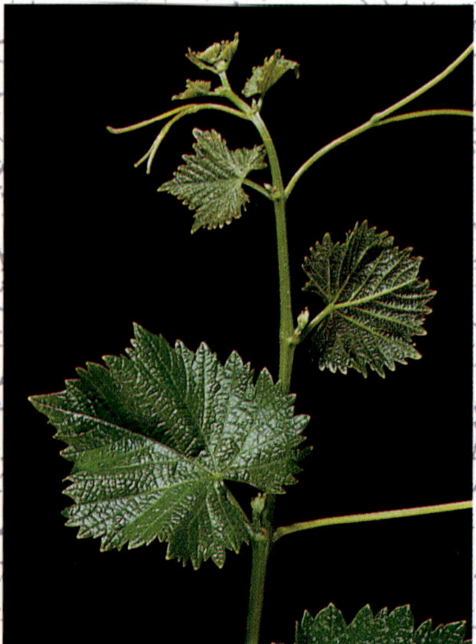

Emerald Riesling

Emerald Riesling is a variety bred by Prof. H.P. Olmo at the University of California, Davis in 1936 and released in 1948. It is a cross of Muscadelle and Riesling. There were 1121 ha of this variety planted in California in 1982 but there have been no new plantings since 1983.

Emerald Riesling is vigorous and high yielding. The young shoot tips are open and free of hairs. Small leaves are a clear mid green with a few erect hairs on the veins on the lower surface. Adult leaves are entire to 3-lobed with a greater depth than width. The superior sinuses are closed and overlapping with the petiolar sinus an open V. The upper leaf surface is lightly bullate and puckered with a few tufted hairs on the lower surface. Teeth are large and irregular and the tendrils are long and branched. Bunches are large, conical and loose with a long green peduncle. Berries are medium in size, round and greenish white when mature. The pulp is bright green and not very juicy. The mature canes are light brown, striated and have darker brown nodes.

Emerald Riesling produces a fresh acid wine with a subtle fragrance. It has lost favour in California, however, because of its tendency to oxidise rapidly and it is unlikely that it will be widely planted. A small area has been planted in Israel.

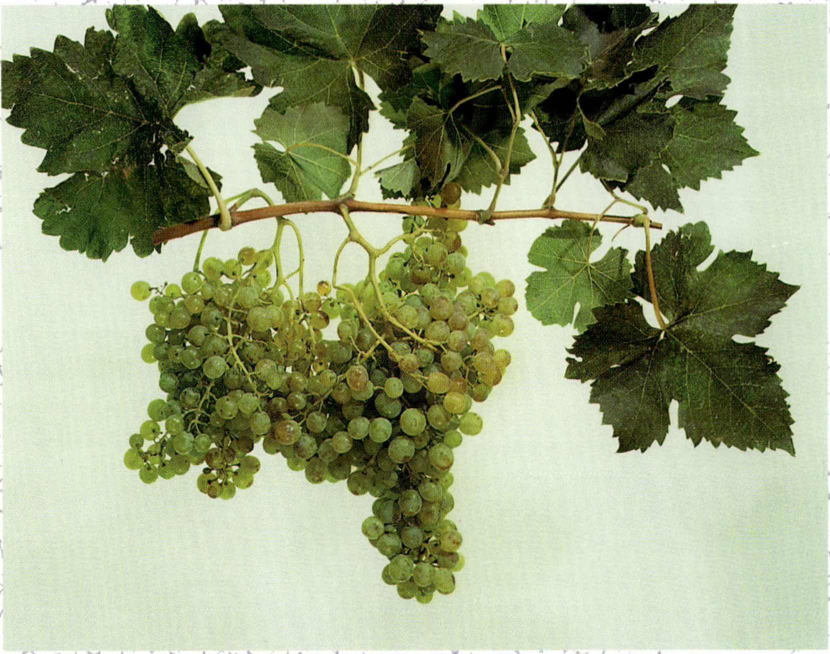
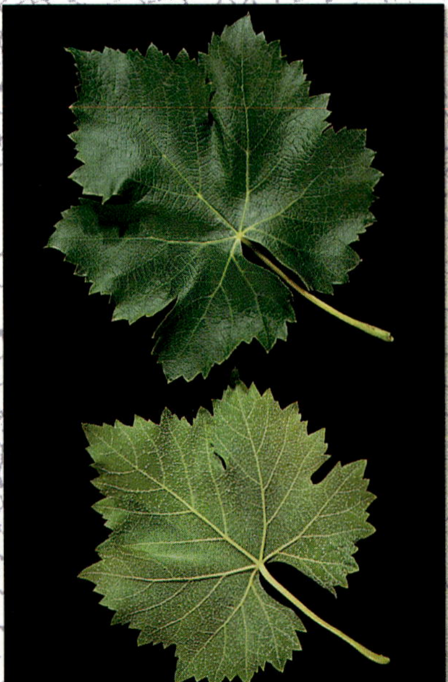
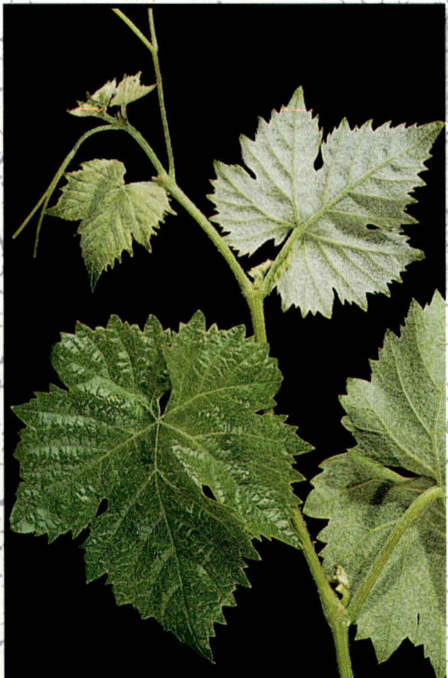

Farana

Farana is the Algerian name for this variety. It is grown in other countries around the Mediterranean Sea, but is known as Planta Pedralba in Spain, Mayorquin in France, Beldi in Tunisia and Damaschino in Sicily. The major plantings were in Algeria and Spain, but with the decline in wine grape plantings in Algeria since independence, Spain would probably have the largest area now. In Australia a little is grown in the Barossa Valley, where it was formerly confused with Trebbiano.

Farana is a vigorous variety with a spreading habit of growth. The leaves are medium to large, 3- to 5-lobed, smooth, with a rather glossy, olive green upper surface and tufted hairs on the lower surface. The petiolar sinus varies from a rather open V to an almost closed lyre in shape. The bunches are large, conical, loose to well filled, with medium, round berries marked with brown where exposed to the sun.

In Spain, Farana is classed with Palomino and Pedro Ximenez as a variety for making dessert wines of high quality. In North Africa it is sometimes used as a table grape, but in the cooler, mountain areas of Algeria it is used in making superior dry white wines.

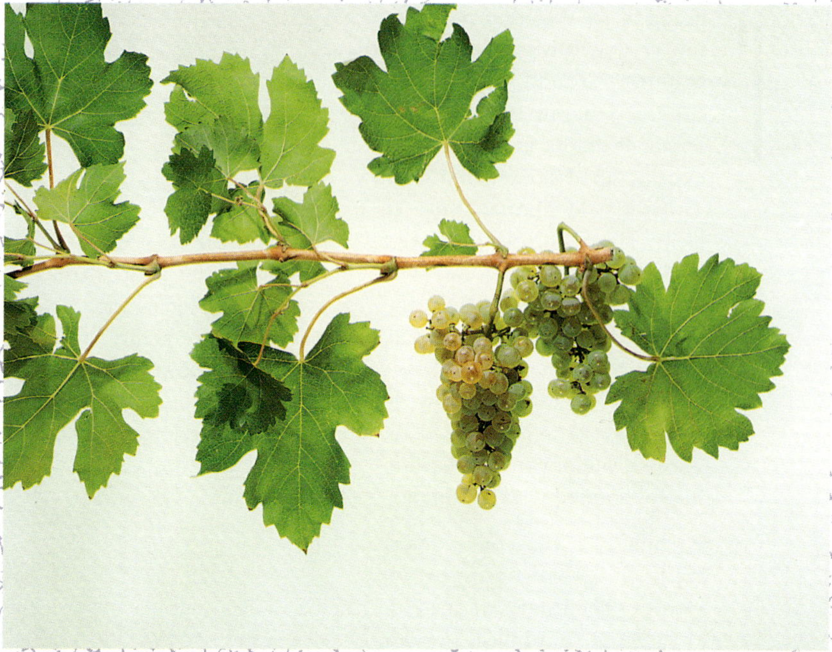

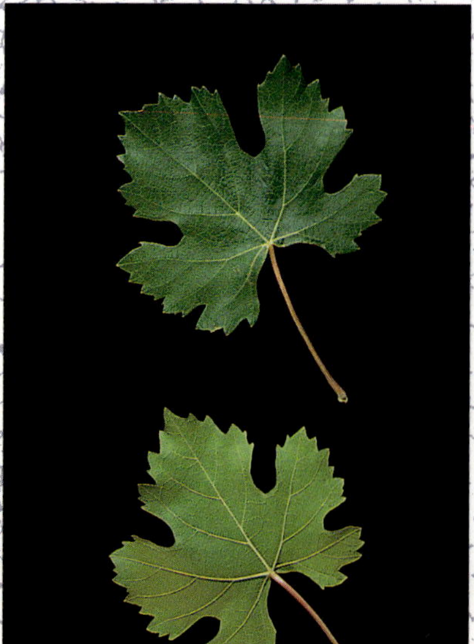

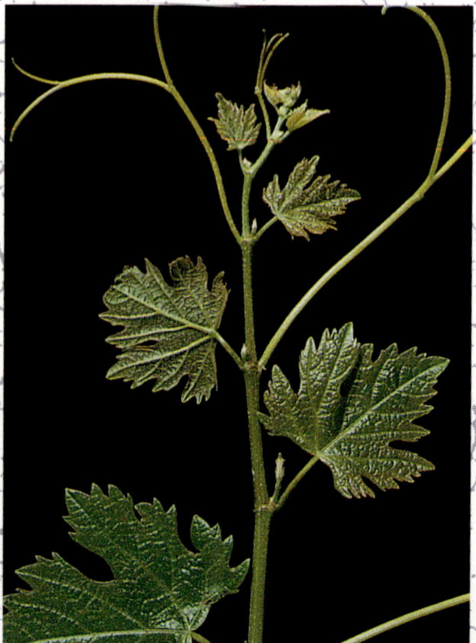

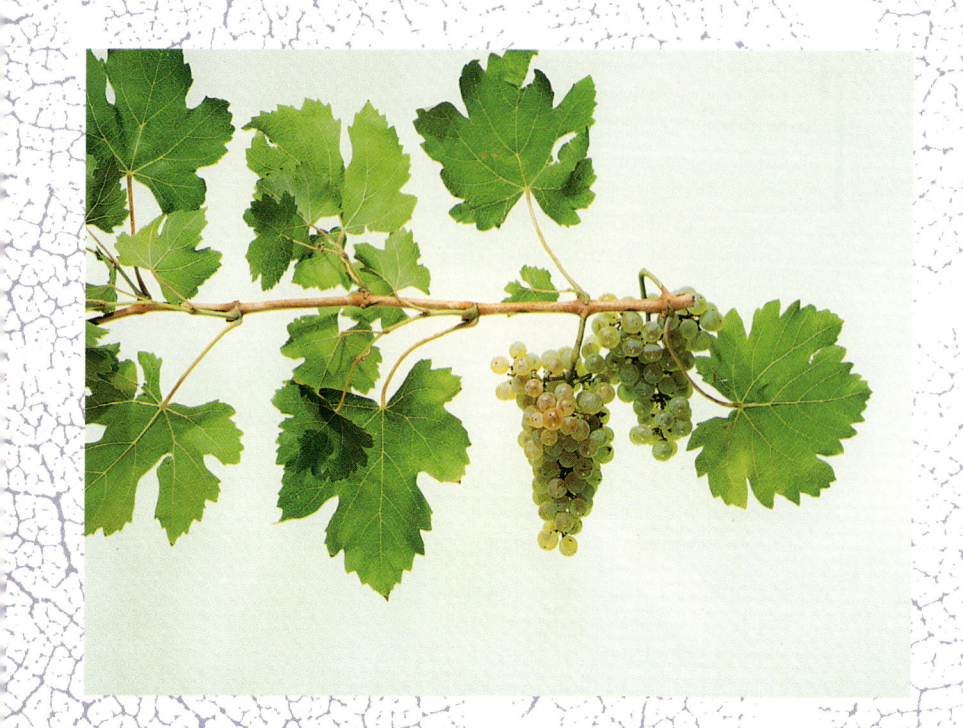

Fetyaska

Fetyaska is an old Rumanian white wine variety which is reported to produce wines of high quality in its country of origin. It is also grown in Bulgaria and in Russia in areas bordering the Black Sea. It is not very fertile in the basal buds and does require cane pruning to obtain adequate yields.

Fetyaska is a vigorous variety but with only average yields. The shoot tip is open and devoid of hairs. The young leaves are dark green with some light bronzing on the tips. Adult leaves are distinctly 5-lobed with open sinuses and glabrous with just a few tufted hairs on the lower surface. The leaf is lightly bullate with some puckering and the petiolar sinus is a wide, open lyre. The cane is light brown and striated with some darkening on the nodes. The bunch is small to medium in size, conical and compact. Berries are small, round and white.

Average yields of about 20 tonnes per hectare are produced in Australia, however the juice composition is well balanced with good titratable acidity levels and low pH. Wine scores from tasting panels have been similar to those from varieties such as Semillon, Trebbiano and Chenin Blanc.

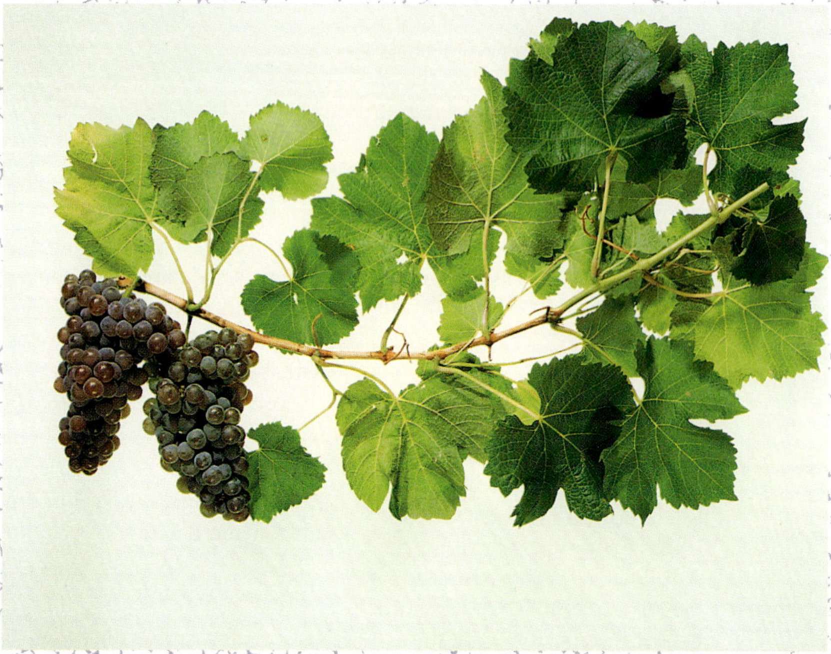

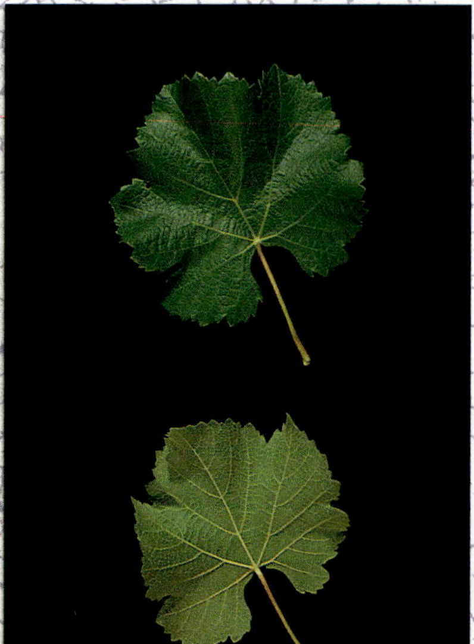

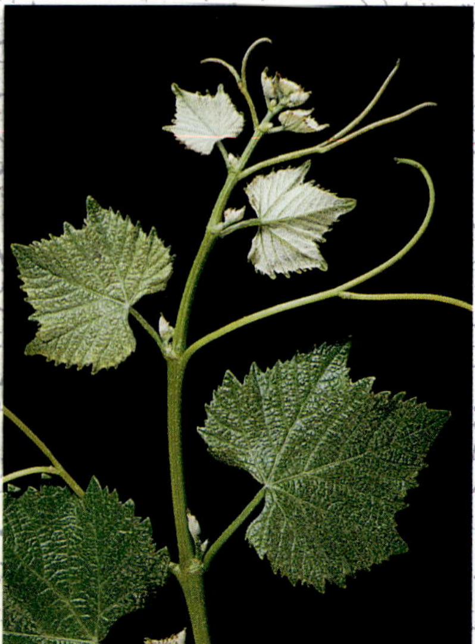

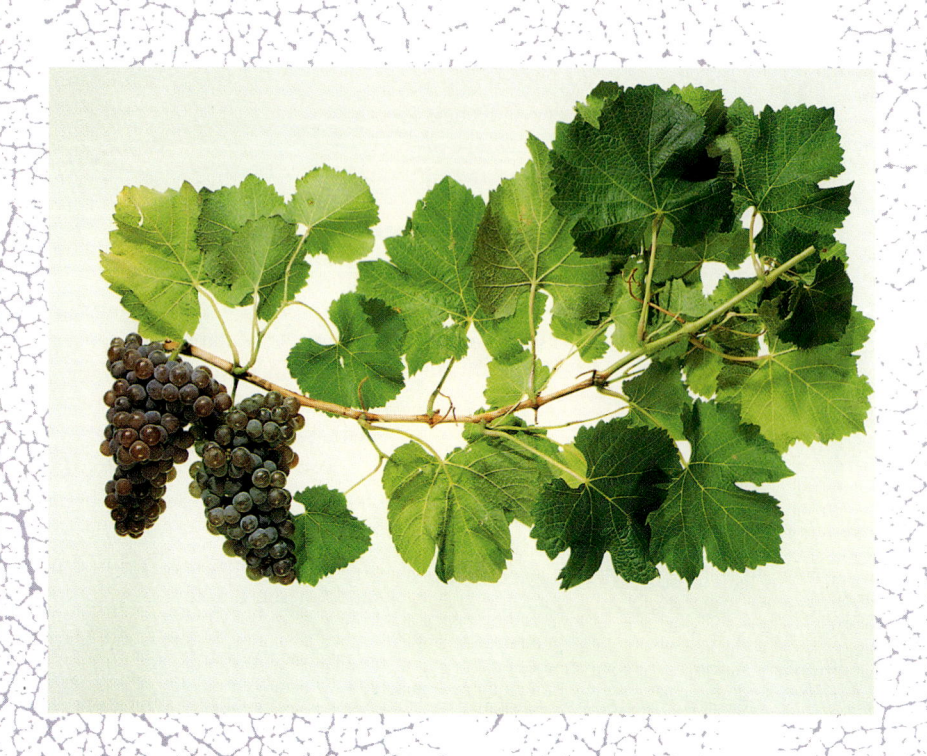

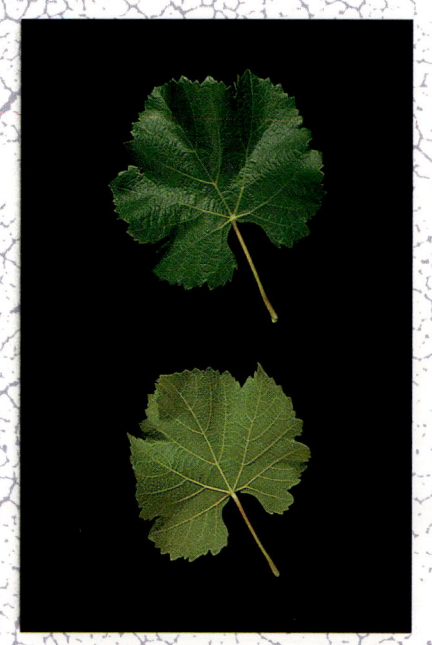

Flora

Flora is a variety bred by Prof. H.P. Olmo of the University of California, Davis in 1938. It was released in 1958 and is a cross of Semillon by Gewürztraminer. It has not been planted to any extent in California or the rest of the world.

Flora has medium vigour. The shoot tip is open, hairy and distinctively white in colour. Juvenile leaves are white on both the lower and upper surface. Young leaves are light green but become dark green and glossy with age. Adult leaves are orbicular and entire to 5-lobed with tufted hairs on the lower surface. The petiolar sinus is an open lyre and the teeth are small and shallow. The leaf is puckered and bullate. Mature canes are light brown with darker areas around the nodes. Bunches are small to medium in size and compact. Berries are small, round and a reddish pink when mature.

Flora ripens in mid February in the warm irrigated regions of Australia. Yields are medium. Berry composition is good with high titratable acidity and low pH. The wines are aromatic and of high quality with a floral bouquet. It ripens in mid to late April in Tasmania.

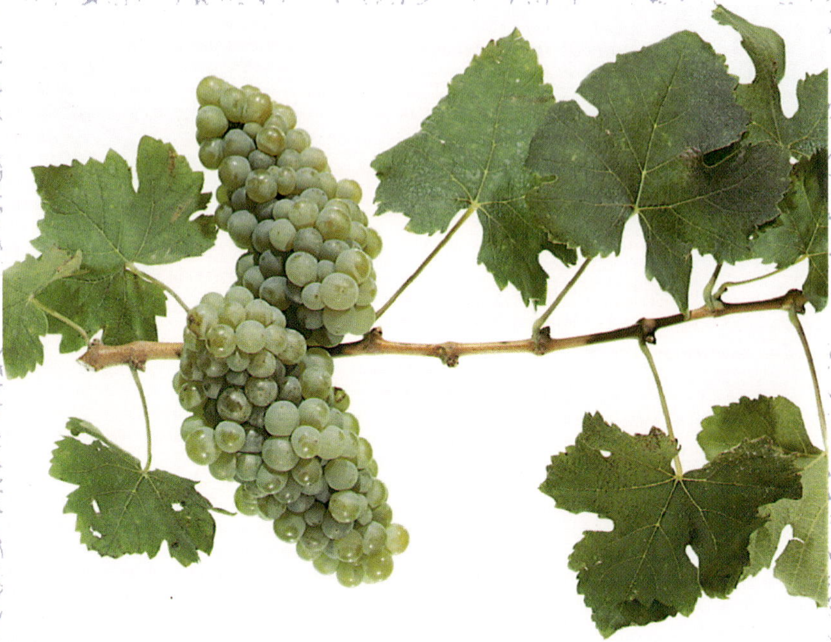

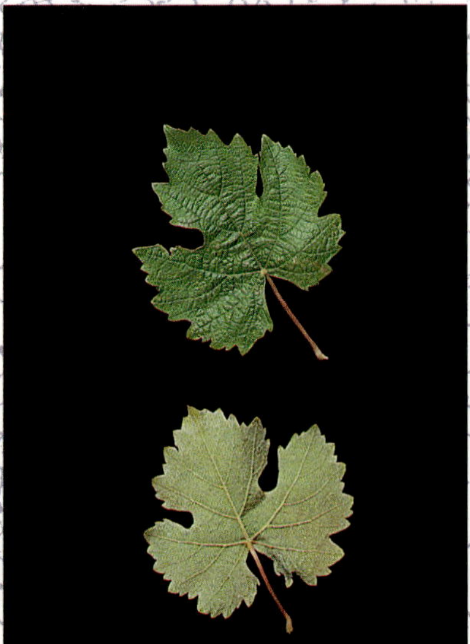

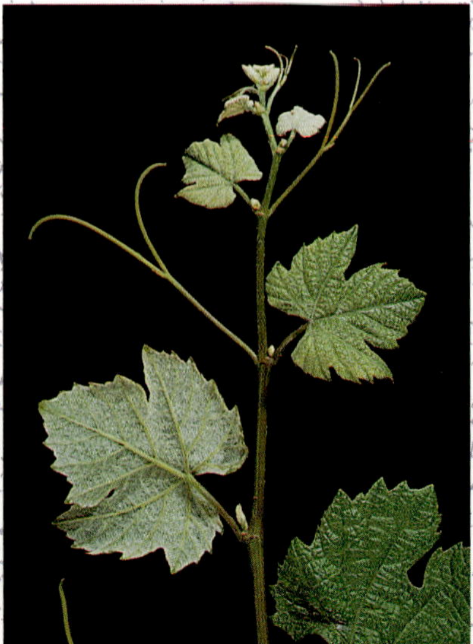

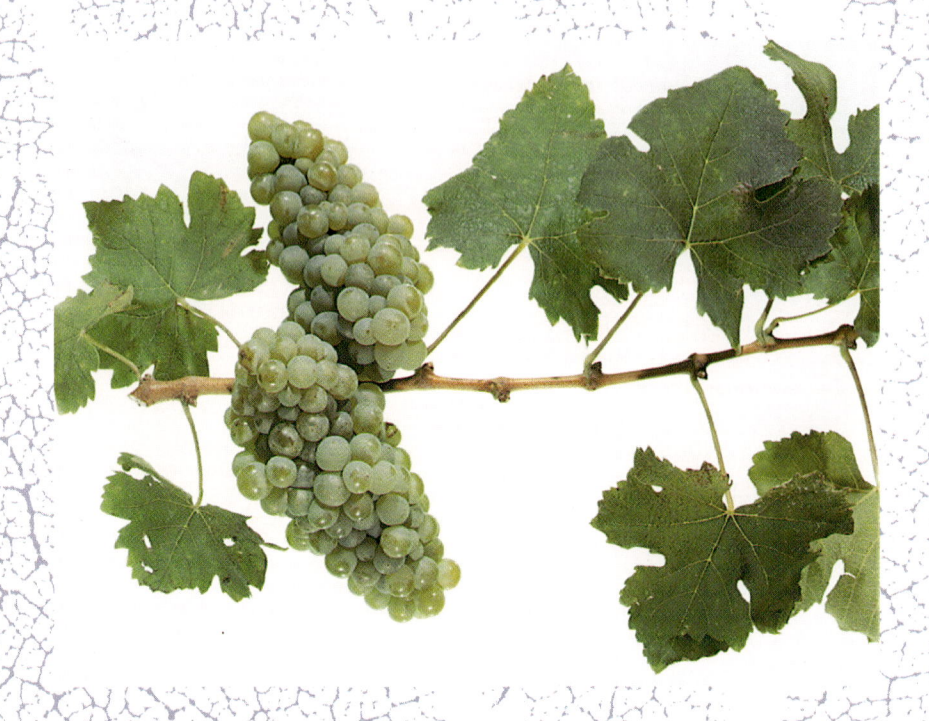

Folle Blanche

Folle Blanche was best known as the variety used in making the brandies of Cognac and Armagnac. The use of rootstocks, which became necessary after the phylloxera invasion, accentuated its susceptibility to Botrytis; its replacement in these areas by Saint Emilion (syn. Trebbiano) and to a lesser extent Colombard is well advanced. This is reflected in the decrease in area in France from 15 800 ha to 3600 ha between 1958 and 1988. However, it is maintaining and even slightly increasing its area in the lower valley of the Loire, where it is used to make a wine exported to Germany for making sparkling wines or a table wine, high in acid and low in alcohol, for which there is some demand.

In California it accumulates enough sugar to make a good table wine and also retains enough acid to be excellent for sparkling wine. However, expansion is restricted by problems with bunch rot and only about 100 ha are grown. The variety has not been grown commercially in Australia. Plantings which might have been Folle Blanche have proved to be the variety known as Sercial in Australia.

Folle Blanche is a moderately vigorous and very productive variety. Even in France yields of 15 tonnes per hectare are not unusual and in Australian irrigation areas yields of 20 to 30 tonnes per hectare should be possible.

The leaves are thick, rough and undulating, 5-lobed with a downy lower surface. The most characteristic feature is the shape of the upper sinuses — like the fingers of a glove. The bunches are medium in size and very compact, with medium berries which are spherical if not pressed out of shape.

The wines of Folle Blanche are usually very acid with a fresh and fruity character. Brandy from Folle Blanche is regarded in France as the best.

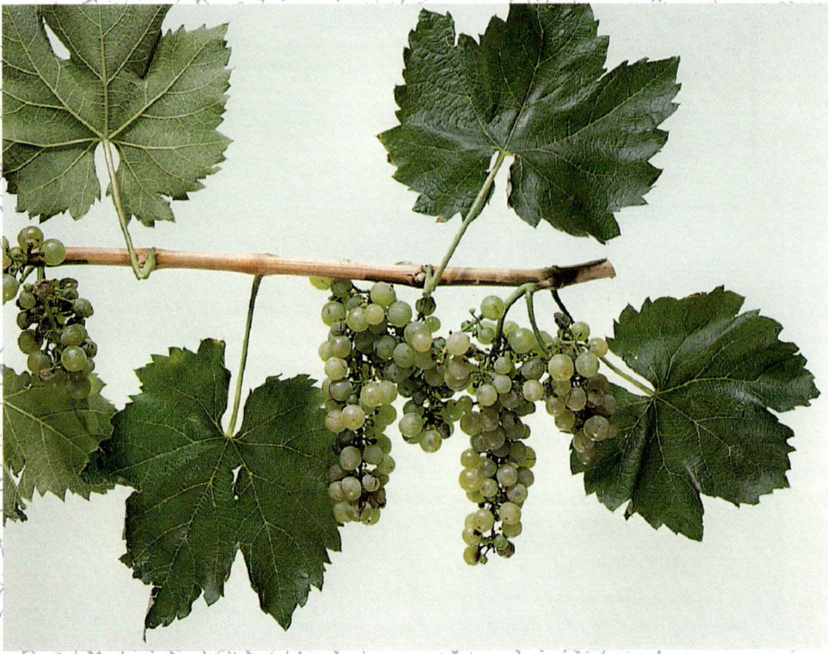

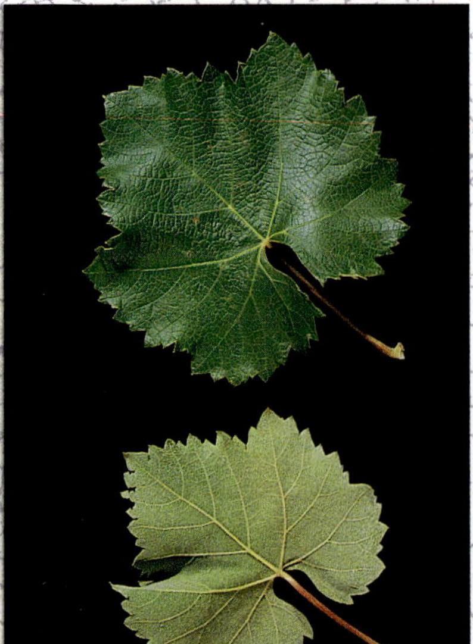

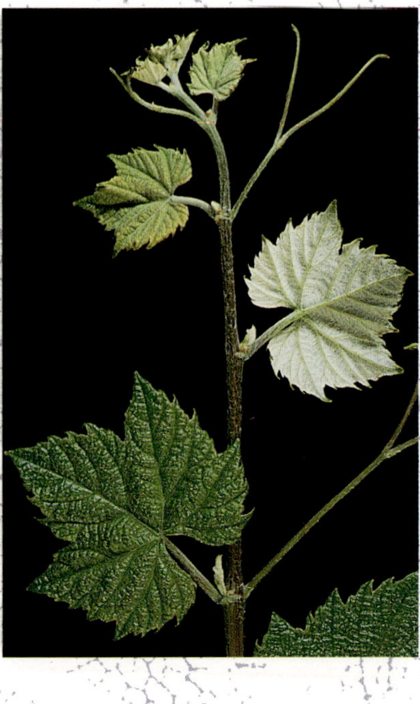

Furmint

Although Furmint is a famous variety, the one major planting in the world appears to be that of about 5000 ha in Hungary for making the special wines of Tokaj-Hegyalja. Furmint has been taken to many other countries but does not seem to have been planted extensively in any of them. It has been in Australia probably since Busby's import of 1832 but has been found only as odd vines in plantings of other varieties, including a mixed planting at Great Western dating from 1868. A variety imported from Italy as Furmint has proved to be another, as yet unidentified, variety.

Furmint is a fairly vigorous variety with a very upright habit of growth. Leaves are medium, entire to slightly 3-lobed, flat with the edge tending to turn inward, smooth to somewhat rough and downy white on the lower surface. The petiolar sinus is lyre shaped. Bunches vary from very small to medium and are loose, cylindrical and winged. Berries are round and are medium in size but show some hen and chicken.

In Hungary Furmint is used in about equal proportions with another variety Harslevelu. The famous Aszu wines are made by a complex process in which the berries dried by noble rot are separated and prepared for sweetening the wine made from the rest of the fruit. These are very fine wines, comparable with the best of the noble rot wines from other countries, but the dry and sweet wines made by simple fermentation in years when there is not enough noble rot for making Aszu wines also show excellent bouquet and flavour

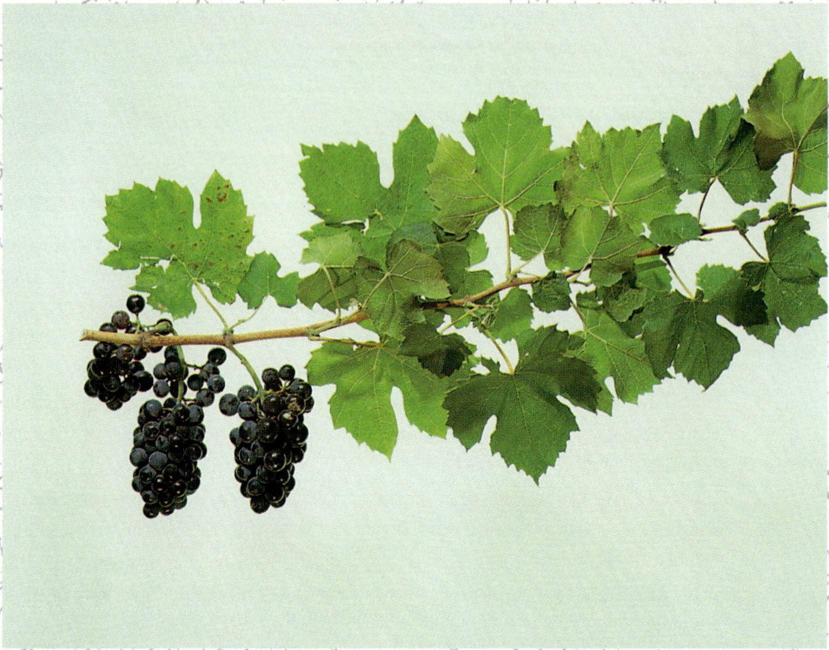

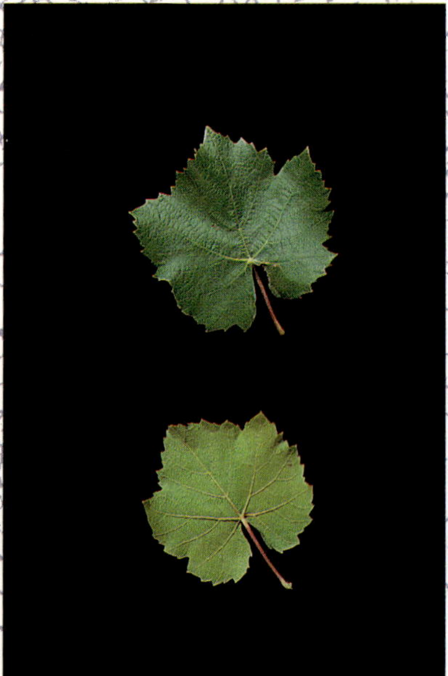

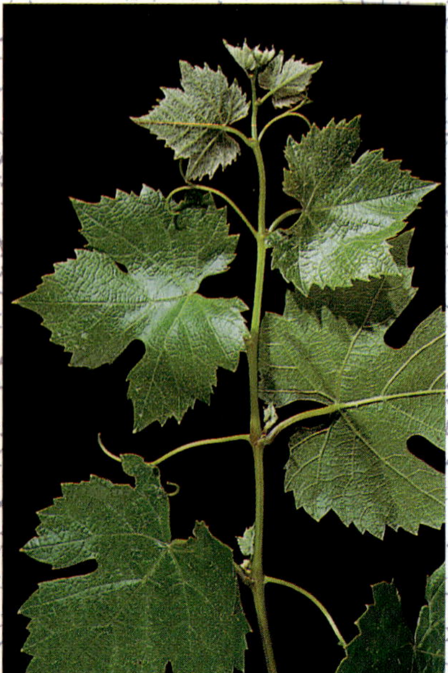

Gamay

Gamay ranks seventh in area, after Carignan, Grenache, Merlot, Cinsaut, Cabernet Sauvignon and Aramon, among the red wine grapes of France, with plantings of about 34 000 ha. A little over half of this is in the Beaujolais region, the rest is scattered through Burgundy and the Loire Valley. Gamay is also grown in the north of Italy and neighbouring areas in Yugoslavia. It has not been grown in California and there has been confusion in Australia because varieties introduced from California as Gamay proved to be wrongly named. The Gamay Beaujolais imported in 1962 was in fact a clone of Pinot Noir with an upright growth habit, and the Napa Gamay imported in 1968 was Valdiguie. Clones of the true Gamay have since been imported from France and a single vine was found in an old planting at Great Western.

Gamay is an extremely fruitful variety and the large amount of fruit carried during formation of the vine appears to inhibit shoot growth to the point where it is difficult to fill a vineyard at spacings commonly used in Australia. It can be pruned to short spurs and will even produce fruitful accessory shoots after a frost. Gamay has medium, plane, smooth leaves, usually almost entire but occasionally deeply lobed, with little or no hair on the lower surface. Bunches are medium, sometimes winged, and compact with medium, slightly oval berries. The fruit ripens early and is susceptible to bunch rot.

Wine from Gamay should be light and fresh, bright red in colour and ready for bottling and drinking without aging. As with Pinot Noir the best wines are likely to come from the cooler areas. The original Gamay variety has fruit with colourless juice. Since 1800, mutants with coloured juice have been selected, first Gamay de Bouze from which in turn came Gamay Freaux and Gamay de Chaudenay. Gamay Freaux has the most intense colour but is reputed to give the poorest quality wine. The others are nearer to the original Gamay in both the colour and quality of their wines.

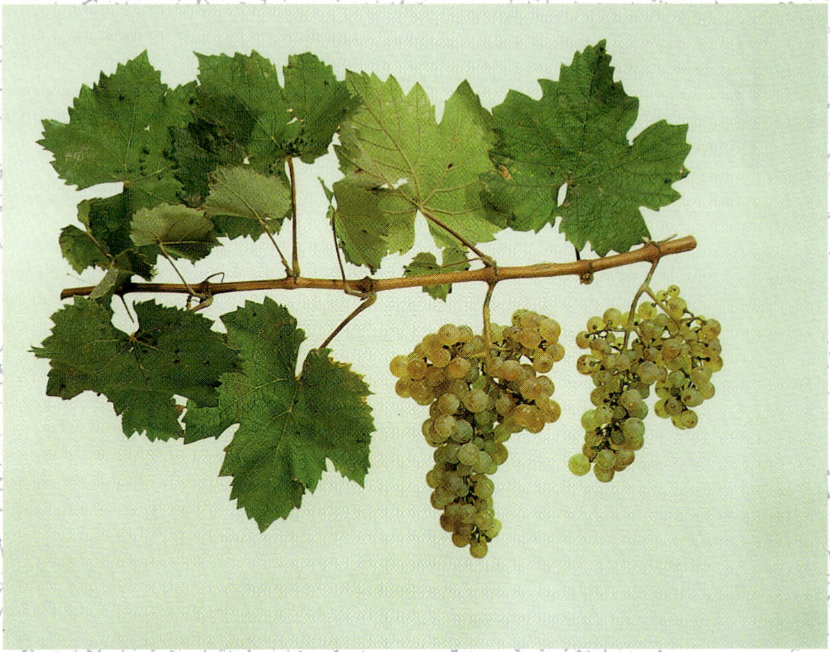

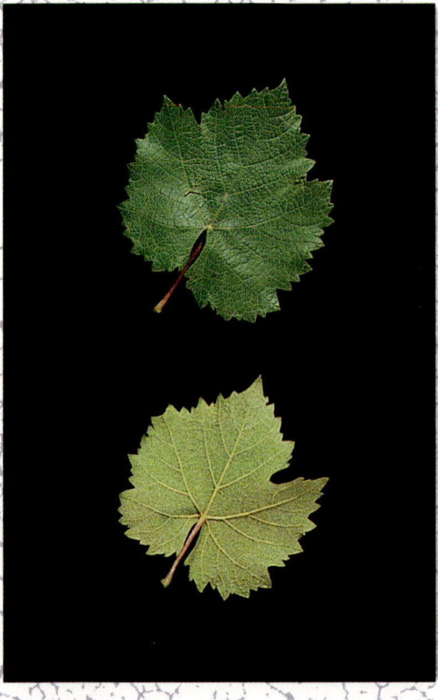

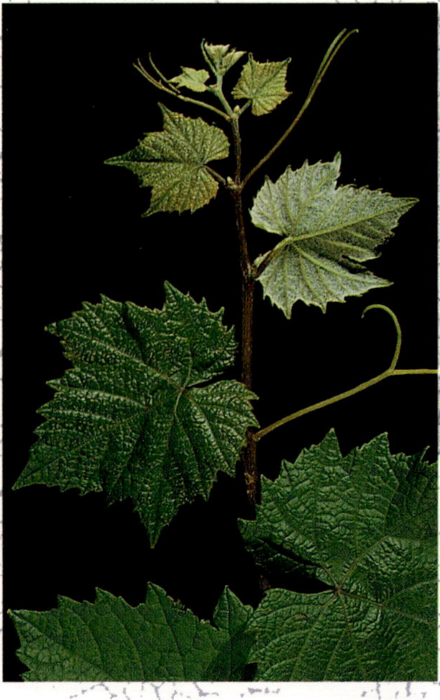

Gouais

Gouais is a minor variety in the centre of France where it is no longer recommended or authorised. It appears to have come to Australia in some of the early introductions in the 19th century, and to have been tried in many areas, but very little remains now. Although there is some resemblance in appearance, Gouais is distinct from Elbling, an important variety in the Moselle areas of Germany and Luxembourg.

Gouais is a variety of moderate vigour and upright growth. It has medium, usually entire, thick, rough leaves tending to fold in about the midrib, with felted hairs on the lower surface. The petiolar sinus is a narrow V with the sides sometimes almost parallel. The petiole is deep red in colour. Bunches are small to medium, winged and well filled, with small to medium, round, golden-yellow berries.

In France, Gouais is regarded as a very inferior variety giving thin, acid, short-lived wines. In warmer areas in Australia, where sugar accumulation might be more satisfactory, the acidity would be no disadvantage, but the variety has had ample time to prove itself and has not become popular.

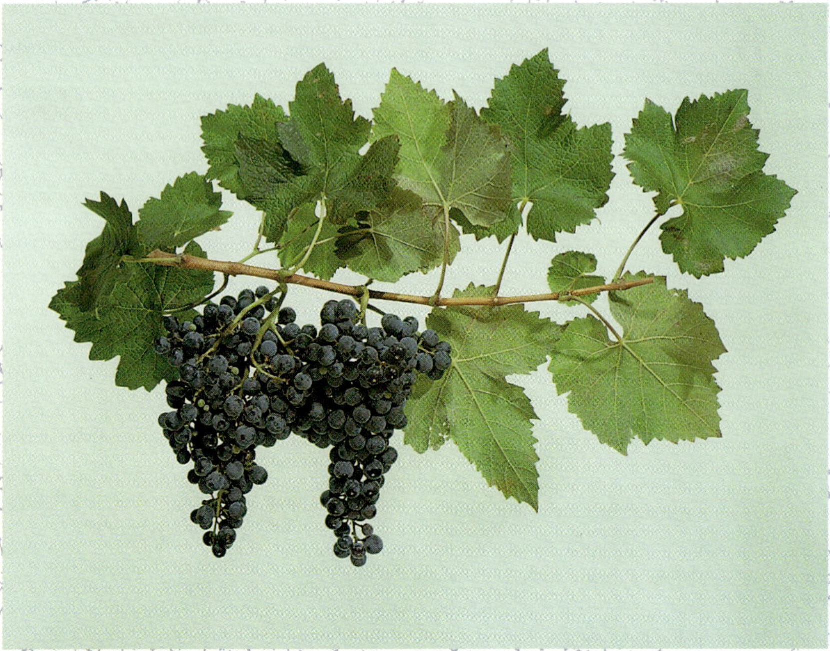

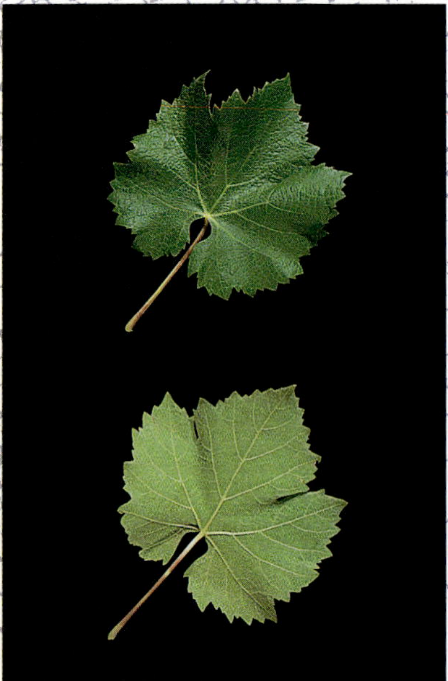

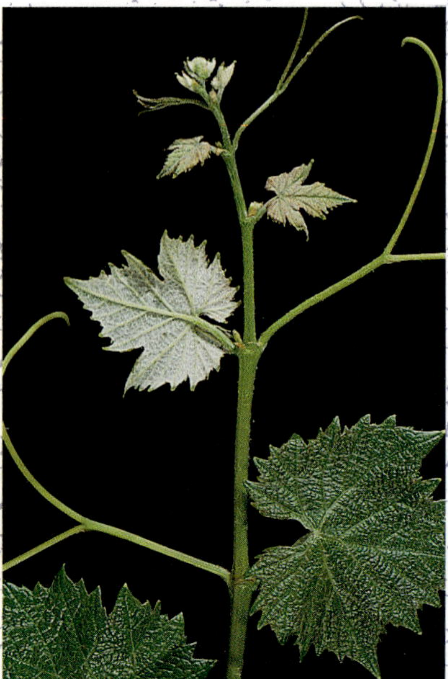

Graciano

Graciano is primarily a Spanish variety. Although it is not one of the major varieties used in bulk wines, there are a few thousand hectares of Graciano in the Ebro Valley, where, along with Carignan and Tempranillo, it is an important component in the appellation wines of Rioja and Navarra. In France the variety is called Morrastel and is recommended in the south, but only about 100 ha remain, possibly because of its relatively low yield. The area of Graciano in Australia is very small. The variety should not be confused with the so-called Morrastel of South Australia, which is really Mataro, nor with the Mourastel imported from California, which is Carignan. In Algeria also, large areas of Mataro were mistakenly called Morrastel. Xeres imported from California proved to be Graciano.

DNA Graciano is a vigorous variety with an upright habit of growth, although not as strongly upright as that of Mataro. Leaves are medium in size, dark green, varying from nearly entire to definitely 5-lobed, with some short hairs on the lower surface, mainly along the veins. The petiolar sinus is more or less closed, sometimes with the edges overlapping. Bunches are medium to large, often winged, compact, with the stalk often becoming woody. Berries are rather small, round, thick skinned and not very juicy. Advantages of the variety are late budburst, which helps to avoid spring frost damage, resistance to drought and relative resistance to powdery mildew.

Wine from Graciano is strongly coloured, rich in tannin and extract, and ages well.

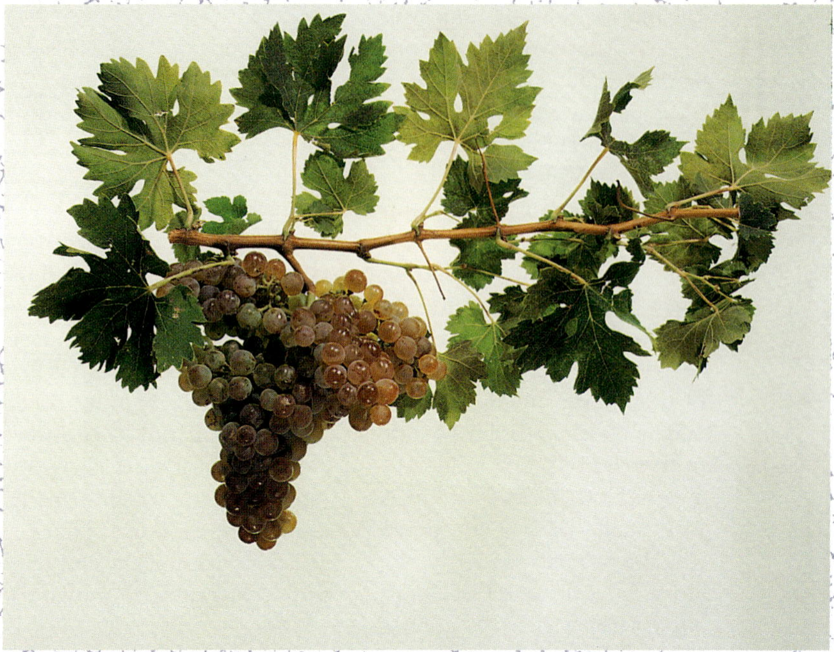

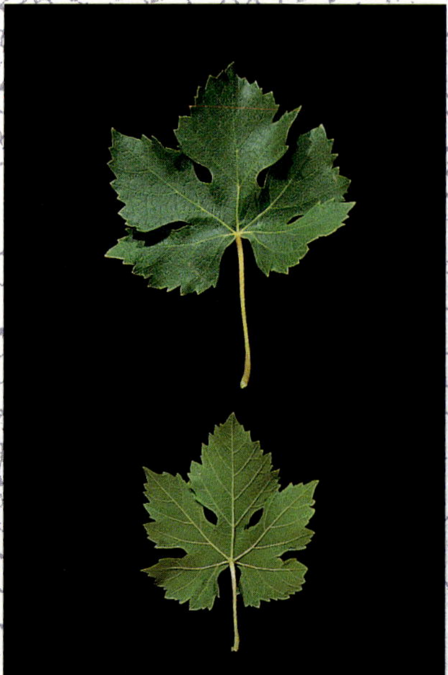

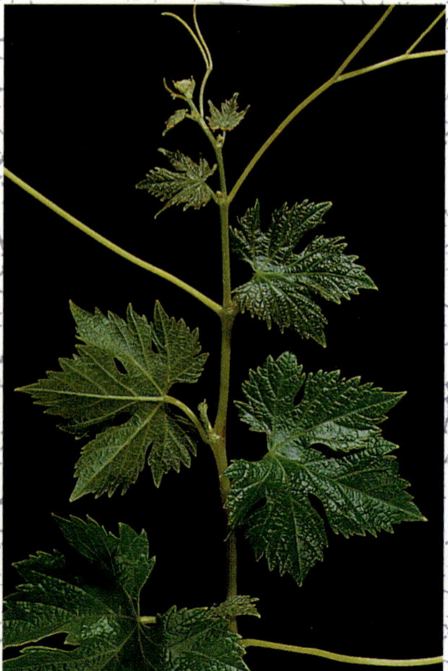

Grec rose

Grec rose was grown in the Mediterranean region of France, both as a table grape and for wine, but has now almost disappeared. It was probably brought to Australia in Busby's collection of 1832, as four names in his list are synonyms of Grec rose. It has been found in variety collections in South Australia and Victoria, where it was earlier known as Wantage or Penarouch and later re-identified as Bermestia. It occurs in an old, mixed planting at Great Western and there could be odd vines in old plantings elsewhere.

Grec rose is a variety of moderate vigour with an upright habit of growth. It has leaves of medium size tending to fold inwards. They are usually deeply 5-lobed, with sharp, narrow teeth, light green, and hairless on the lower surface apart from some bristly hairs on the veins. The petiolar sinus has an open lyre shape. Bunches are medium to large, sometimes very large, conical, shouldered, well-filled to compact, with medium, round berries of a characteristic brownish pink colour.

Grec rose was grown in France because of its large yield. It would be useful in Australia only for bulk wines or distillation.

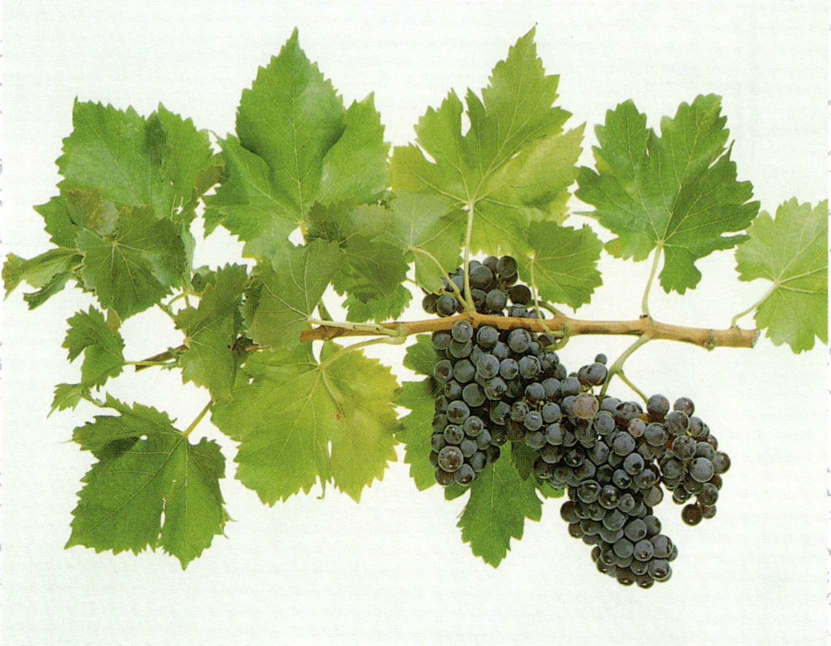

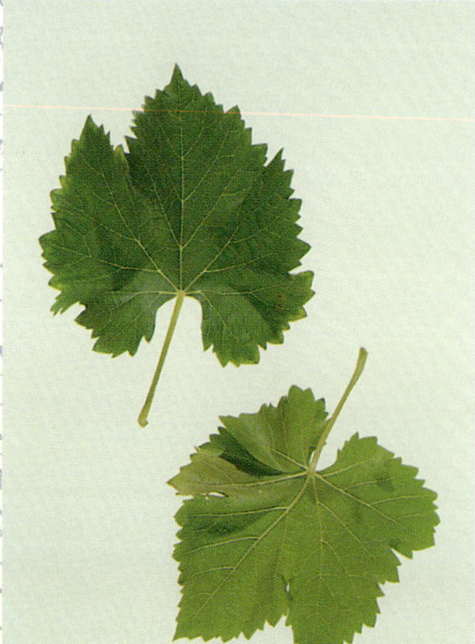

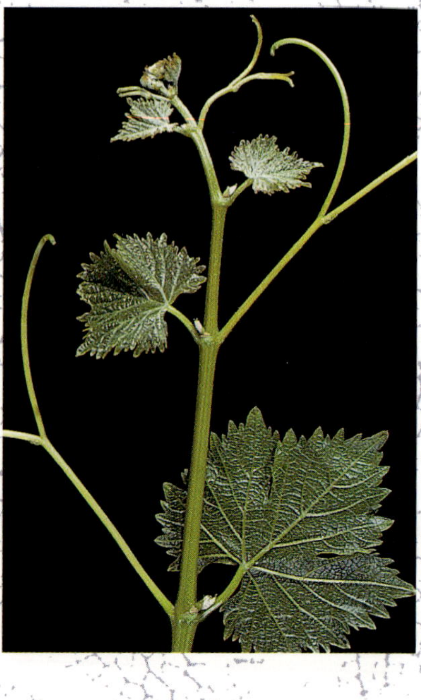

Grenache

Grenache is a very important variety in southern Europe. There are about 105 000 ha in Spain, where it is known as Garnacha. In France the area of Grenache increased from about 52 000 ha to 87 000 ha between 1968 and 1988, and has no doubt continued to increase, mainly at the expense of the high yielding but poor quality variety Aramon. It is an important variety on the island of Sardinia, where it is called Cannonao. Grenache is also present in Sicily and the southern Italian mainland under the names of Granaccia and Alicante. Other than Europe and what may remain in Algeria, the only appreciable areas are in California and Australia. California has 5200 ha. In Australia there are 1900 ha, 1700 ha in South Australia and the rest fairly evenly divided between New South Wales, Victoria and Western Australia.

Grenache is an upright growing and potentially vigorous variety, but the amount of growth is more sensitive to crop level than for most common varieties. It has medium, smooth, shiny, clear green, 3-lobed leaves which are hairless above and below. Bunches are medium, short and broad, compact with medium, round to short oval berries. The berries have a heavy bloom and can vary in colour from reddish pink to blue black according to crop level. Experience in Algeria as well as Australia has shown that Grenache is one of the best varieties in resisting hot, desiccating conditions.

The wines of Grenache are low in colour by Australian standards and age rapidly. Nevertheless Grenache is regarded as a premium variety in France if it is not cropped too heavily. It is used alone only in rosé and fortified wines. For red table wines it is usually combined with varieties such as Carignan and Mataro which provide acid and tannin, and Cinsaut which gives smoothness. Shiraz, Clairette and other varieties may also be included to increase complexity.

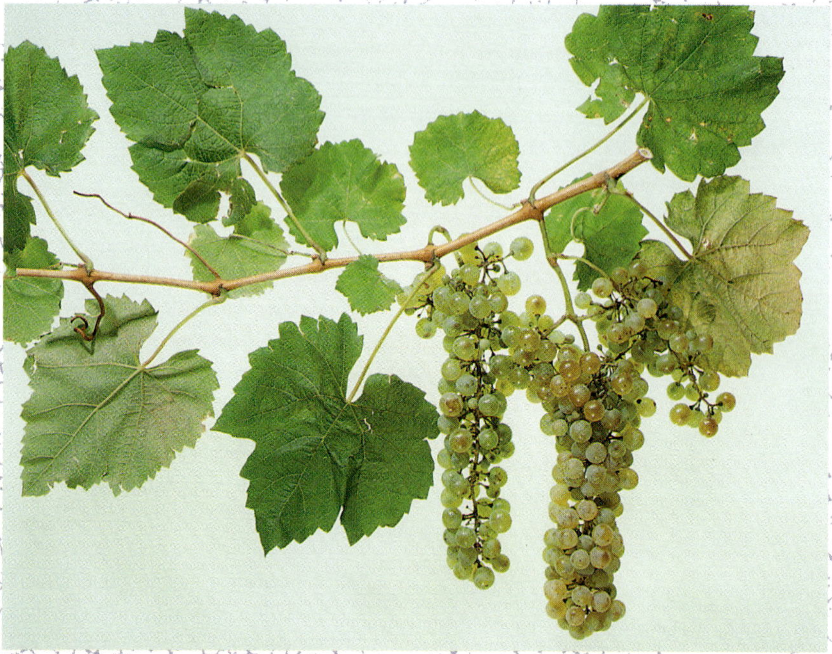

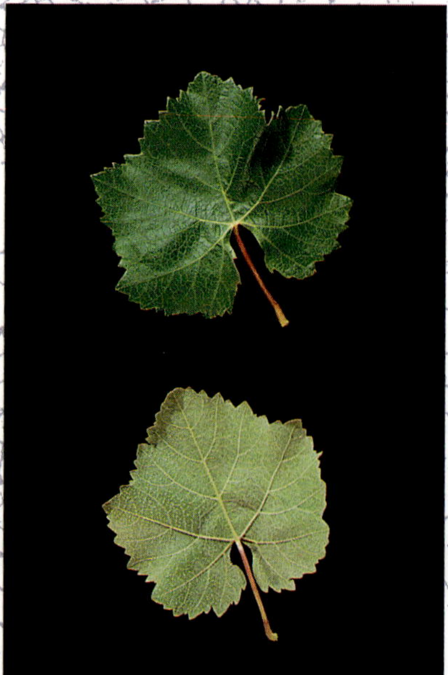

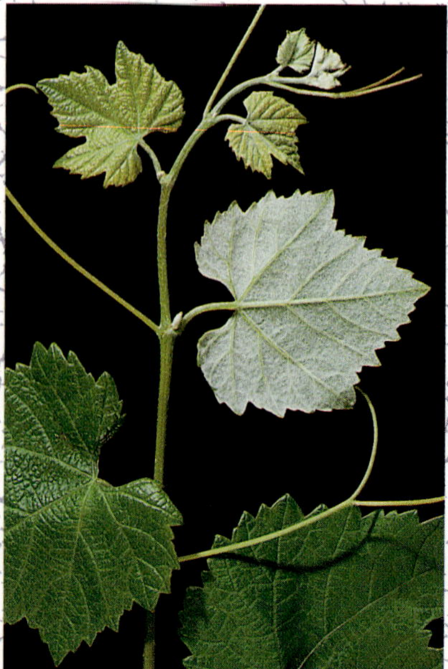

Harslevelu

Harslevelu is an old variety from the Tokay region of Hungary where it is blended with Furmint to make the famous sweet wines of the region. Both of these varieties seem well suited to the concentrating effect of *Botrytis cinerea* or 'noble rot' and produce wines which more than rival the wines from the Sauternes region of France.

Harslevelu is a vigorous variety which has an early budburst and ripens a little before Furmint. Spur pruning is normally used as cane pruning can result in small bunches and lighter style wines. It has an erect growth habit. The opening buds are large and cottony. The young shoot tips are open, white and hairy with the young leaves a clear green on the upper surface. The growing shoots are long, large, straight, striated and a clear yellow in colour. Tendrils are numerous, strong and branched. The young leaves are heavily felted on the lower surface. Adult leaves are large, wider than long and almost orbicular with a petiolar sinus in the shape of a closed lyre. They are entire to 3-lobed, dark green on the upper surface with felted hair on the lower. Teeth are small.

Bunches are large to very large, long, winged and divided with a long, clear green peduncle. Berries are round, medium in size and a golden, transparent yellow colour at maturity. They have a thin skin and the juice is sweet and aromatic.

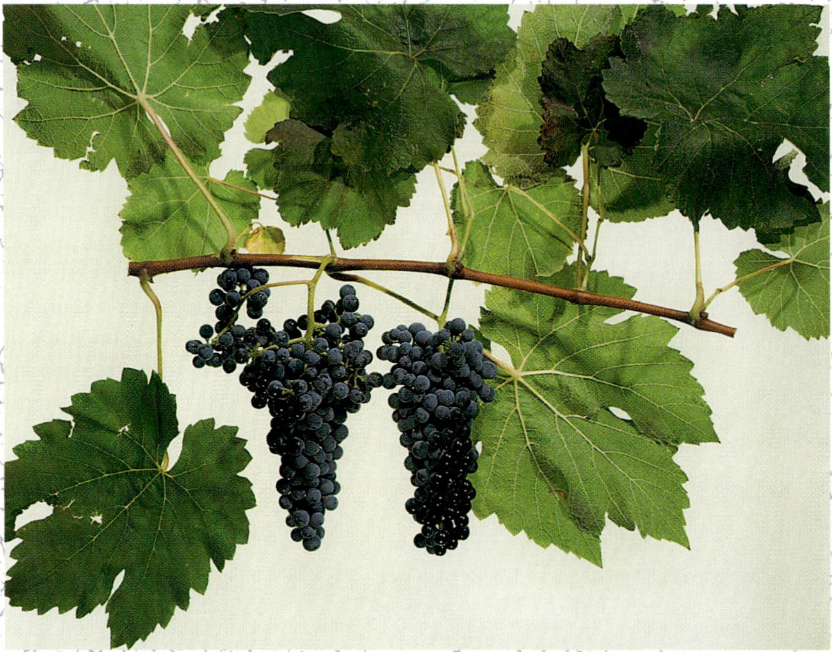

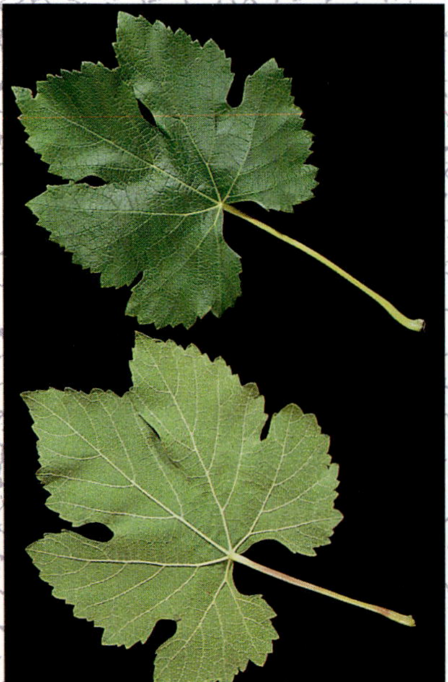

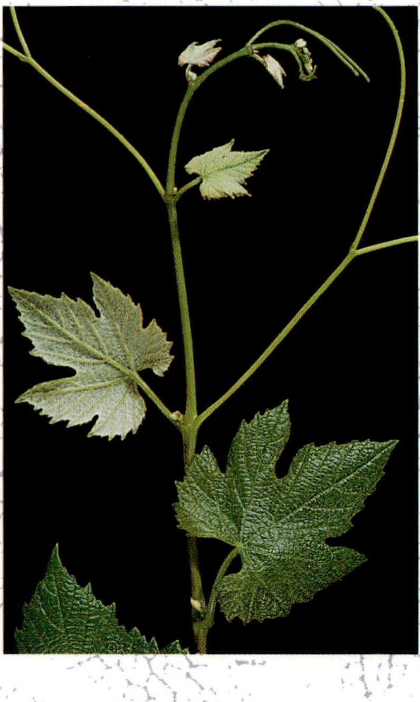

Jacquez

Jacquez is one of a group of varieties to which the species name *Vitis bourquiniana* or *V. bourquina* has been given. The original members, including Jacquez, appear to be natural hybrids between *V. aestivalis* and *V. vinifera*. Jacquez, generally known as LeNoir in the United States, is thought to have originated in Georgia or the Carolinas but has not been widely grown in that country. It became very popular in France after the vineyards were devastated by phylloxera, but was banned from the production of wine for sale about 50 years ago. It may still be used as a rootstock and remains popular for this purpose on suitable soils in South Africa. In Australia, where it has been called Troya, annual production has been about 100 tonnes, mostly in the Murrumbidgee Irrigation Area with a little in the Hunter Valley.

Jacquez is a vigorous variety which has inherited resistance to fungal diseases from *V. aestivalis*. It is practically immune to powdery mildew. It has very large leaves which are 3- to 5-lobed, smooth and rather glossy, lighter in colour below than above, with fine, cobwebby hairs on the lower surface. Bunches are medium to large, well shouldered and well filled, with rather small, round, intensely coloured berries which have a heavy bloom and coloured juice. Canes are a deep reddish brown in colour.

Wine from Jacquez has a deep colour and a strong, unusual flavour less unpleasant than those of the hybrids of *V. labrusca*. The variety might also be useful for producing an unfermented grape juice resembling blackcurrant juice.

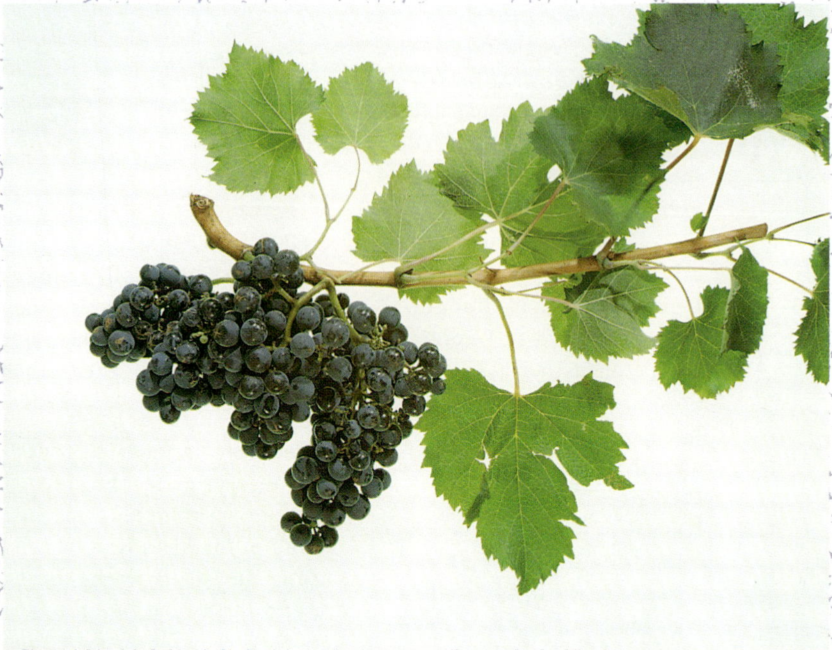

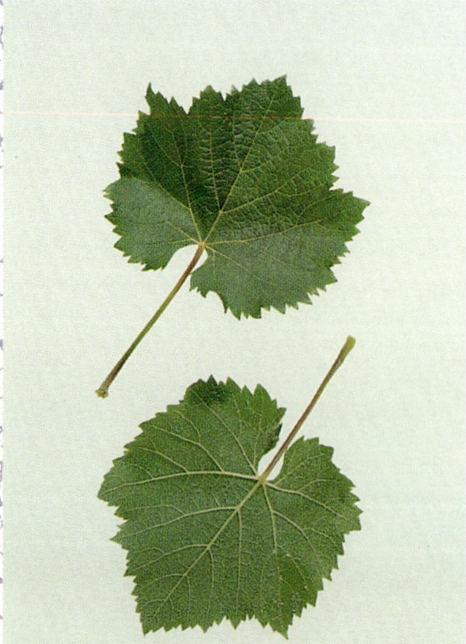

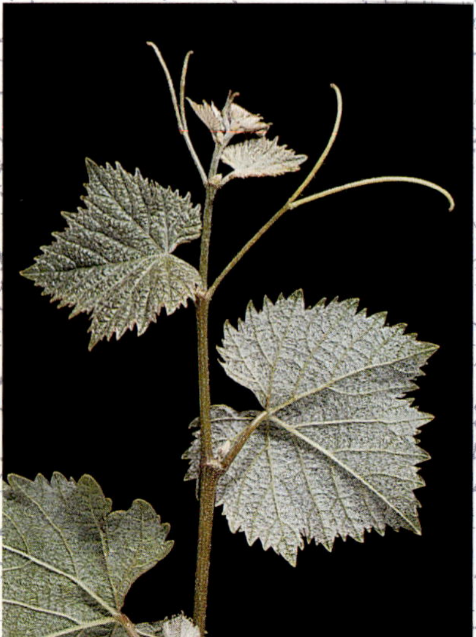

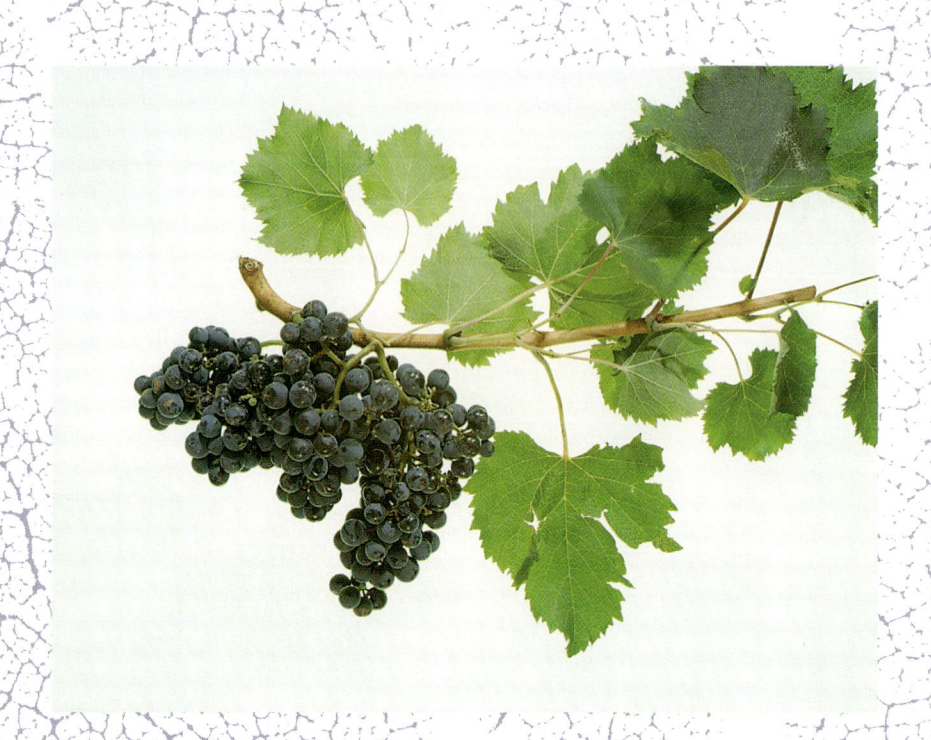

Malbec

The official French name for this variety is Cot. Malbec, sometimes spelt Malbeck, is an approved synonym. Another synonym used in some areas is Auxerrois, and the white variety of this name grown in the north east of France may be so called because its leaves are very similar to those of Malbec. In 1988 there were 5280 ha of Malbec in France, compared with 10 750 ha in 1958. About half is in Gironde, most of the rest in the departements to the east of Bordeaux, and some in the Loire Valley. It is permitted in wines of controlled appellation in all these areas and is the principal variety in Cahors. It does not seem to have spread elsewhere in Europe and there are less than 45 ha in California, but it is one of the most important black varieties in Argentina with 10 000 ha. There are 6000 ha in Chile. It is not clear how much Malbec there is in Australia as at least two other varieties, Dolcetto and Tinta Amarella, have sometimes been incorrectly called Malbec. There would seem to be more than 300 ha of true Malbec, much of which has been planted since 1970.

Malbec is a vigorous variety with a rather spreading habit of growth. It has medium, rough, dark green leaves, usually 3-lobed but sometimes entire or 5-lobed, tending to roll back at the edges, with scattered tufts of hair on the lower surface. The open V of the petiolar sinus is characteristic. The bunches are medium, well filled if set is favourable, with rather small, round berries. The variety has the potential for yielding well and can tolerate rain at harvest but has the defect of setting very poorly in some seasons. This problem is accentuated by the application of nitrogenous fertilisers and by the choice of rootstocks which readily take up nitrate from the soil. Malbec can be pruned to either canes or spurs and as the berries are not very firmly attached mechanical harvesting should be possible with little damage.

With moderate yields in cooler areas Malbec makes a balanced wine of good colour which has a less intense varietal aroma and is softer than Cabernet wines. It combines well with the other Bordeaux varieties to give wines designed for earlier maturity rather than very long holding. In the Loire Valley it is sometimes used alone to give rosé wines with strong fruit character; in Cahors, on the other hand, it is used with up to 30% of Semillon or three other minor varieties to give a very deep red wine which is aged for several years before bottling. When grown for high yields in hot areas, it can be used in the same way as Cinsaut.

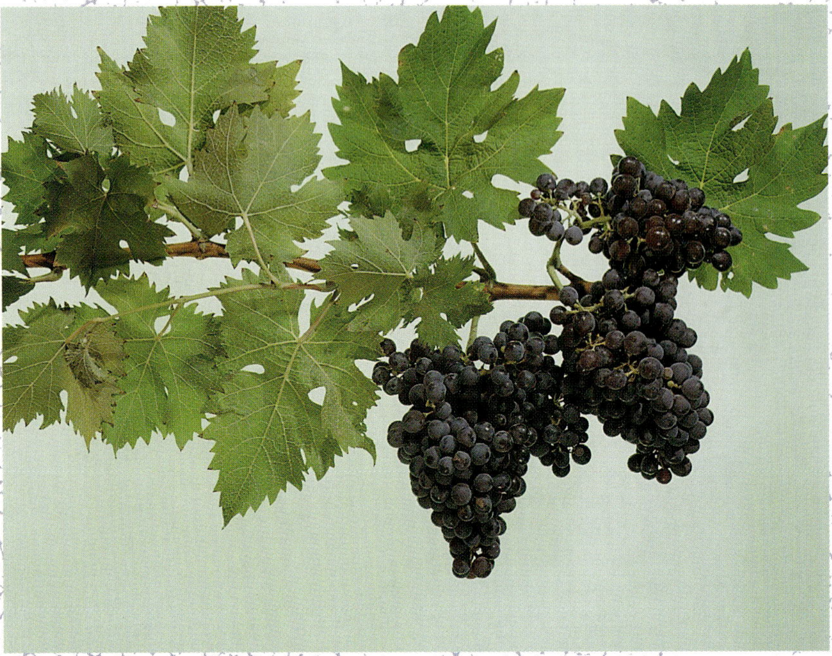

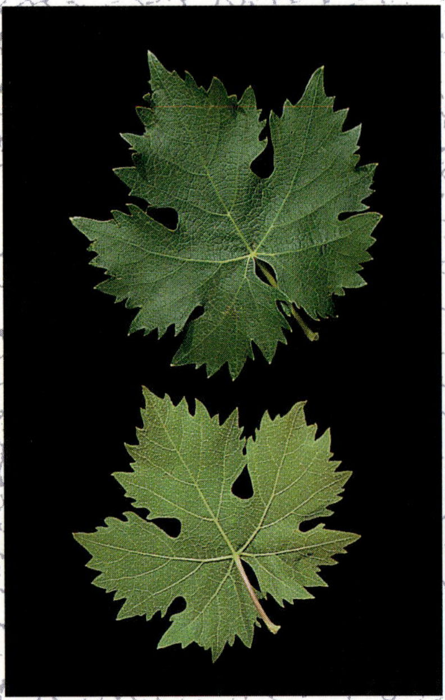

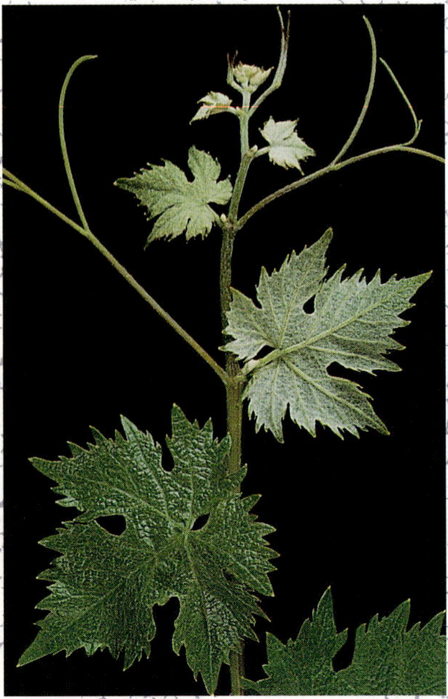

Mammolo

Mammolo is a minor variety from the Tuscany region of Italy, where it was traditionally used as one of the accessory varieties in Chianti. However, it was not included in the authorised varieties for this wine when the regulations for controlled appellation were drawn up in 1967 and may decline in importance. It does not appear to have become important anywhere else in the world and in Australia it was found only as odd vines at Mudgee. Some small plantings may have been made in Australia as five tonnes of Mammolo were recorded as being crushed in 1981.

Mammolo is a vigorous variety with an upright habit of growth. It has leaves of medium size which are 5-lobed with very deep upper sinuses, and shallower lower sinuses and which have tufted hairs on the lower surface. The leaves fold strongly inwards making the petiolar sinus, which is in the shape of a narrow lyre, appear to have overlapping edges. Bunches are medium, conical and compact, and berries are rather small and irregular in size, round and not deeply coloured.

Mammolo owes its name to the fact that its wines on aging develop a bouquet resembling the scent of violets, for which the Italian name is mammola. In Italy, Mammolo is generally not fermented alone but in conjunction with other varieties so that it will add its particular bouquet to the wine.

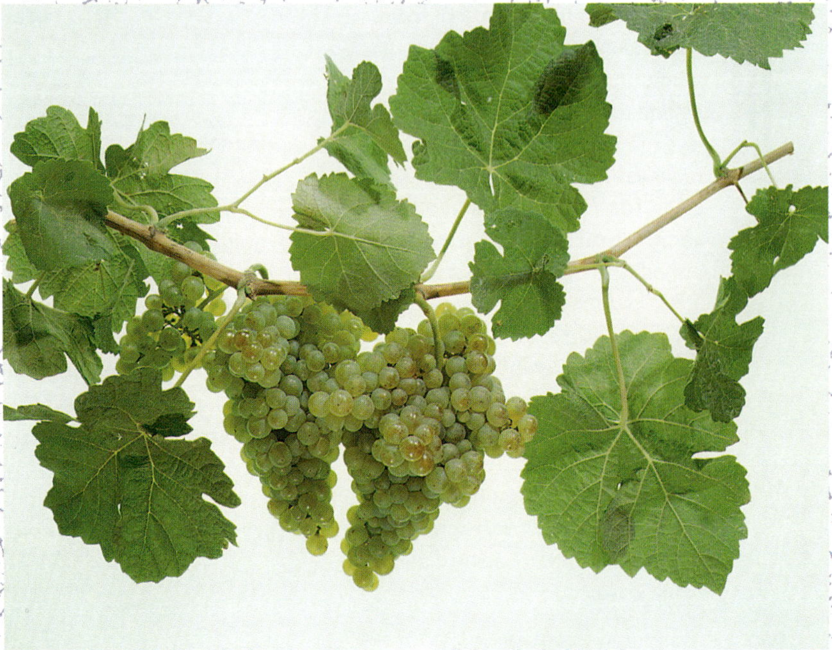
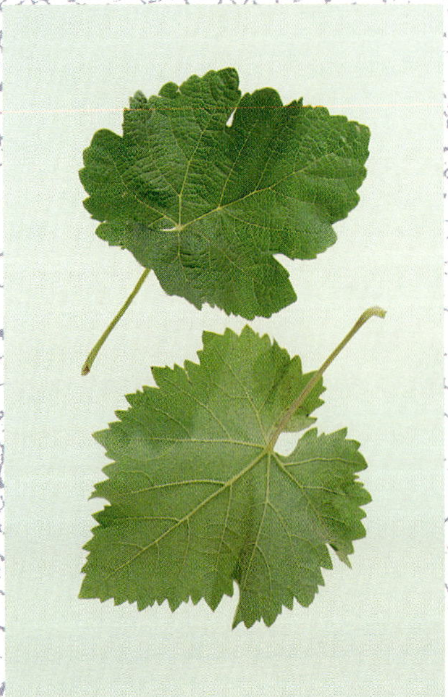
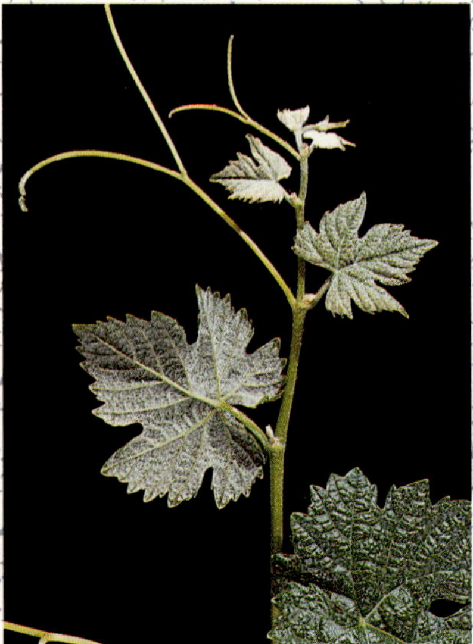

Marsanne

Marsanne is a minor French variety from the Hermitage area in the Rhone Valley. Plantings in France decreased from 282 ha in 1968 to 253 ha in 1979 but increased again to 406 ha by 1988. There is also a little of the variety in the Valais in Switzerland under the name of Ermitage. About 44 ha are planted in Australia, by far the greater part in the Goulburn Valley and north-eastern Victoria and the rest in New South Wales.

Marsanne is a vigorous and productive variety with a spreading habit of growth. It has very rough, dark green leaves, usually 3-lobed with a considerable overlap at the petiolar sinus, and tufted hairs on the lower surface. Bunches are medium, conical and well filled, with small, round berries which become brownish gold when ripe.

In France, Marsanne used alone has the reputation of producing light wines with little varietal character which age very quickly. The addition of a proportion of another variety, Roussanne, is considered to improve quality. Wines from Marsanne in Australia appear to have more body and character, but the major plantings have an appreciable proportion of other varieties present which may be taking the place of the Roussanne.

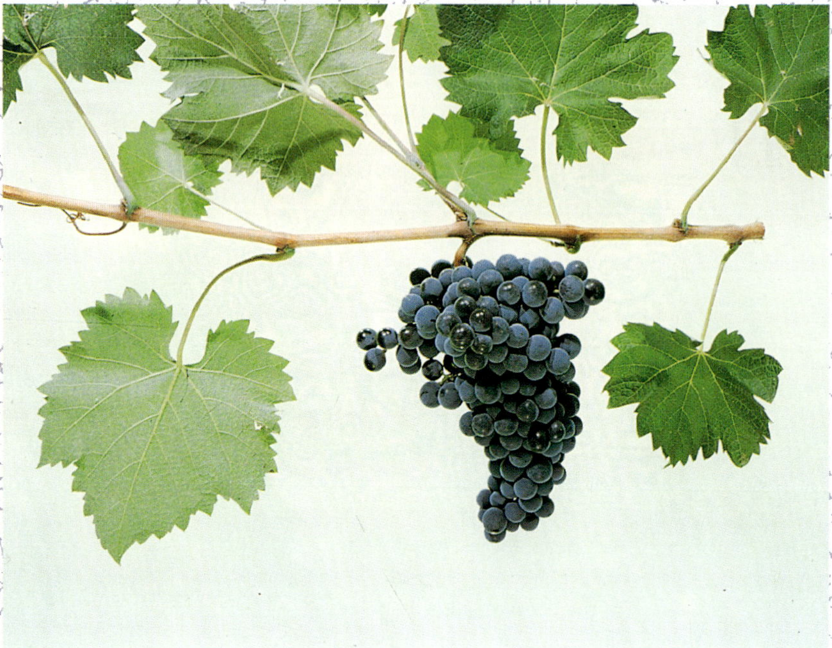

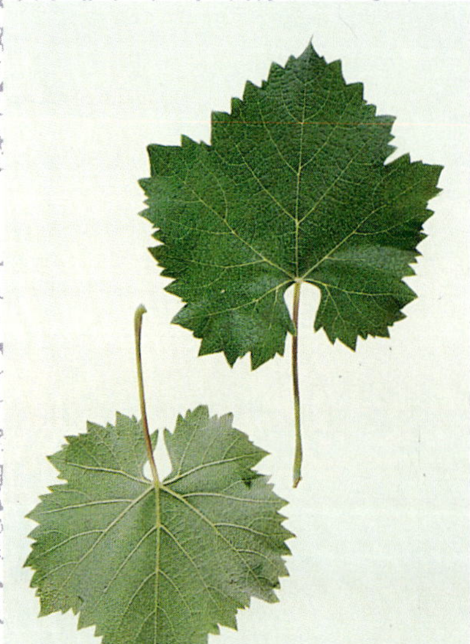

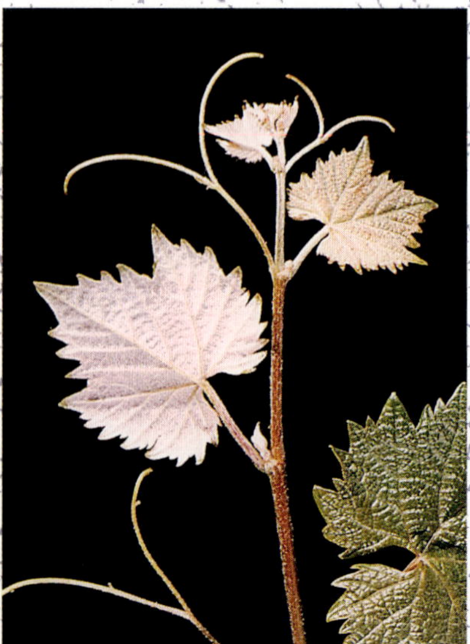

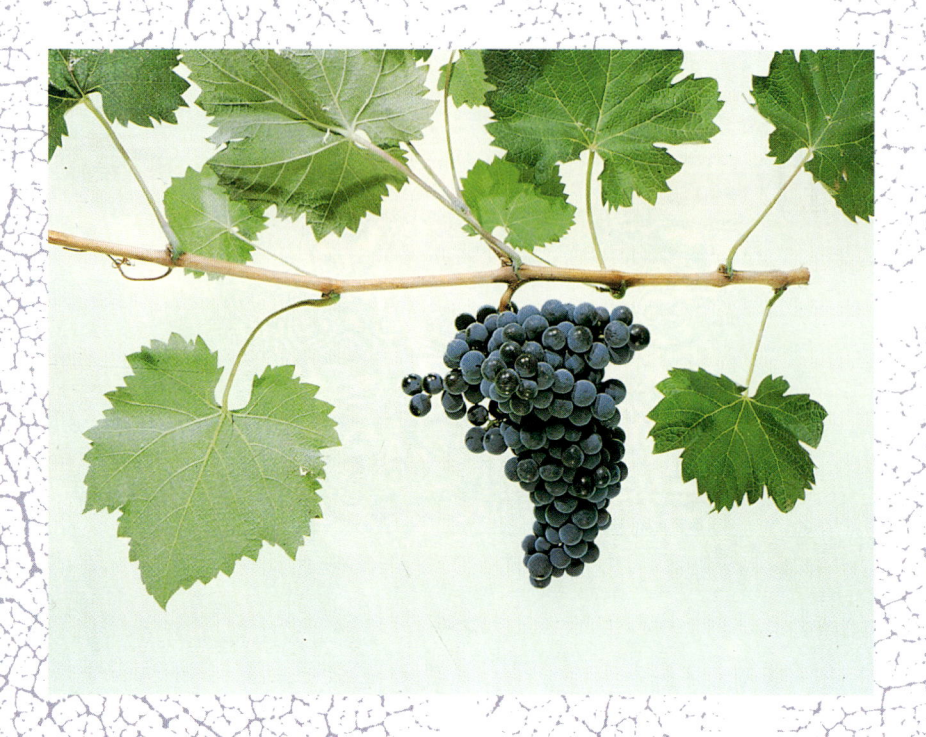

Mataro

Mataro is another variety from southern Europe and appears to need more warmth than Grenache. In Spain, where it is also called Monastrell or Morastell, there are 110 000 ha. In France the area planted has increased from 858 ha in 1968 to 5609 ha in 1988. Apart from small plantings under the name of Balzac in the Cognac area where the fruit does not ripen properly, it is confined to the warmest areas of Provence, where it is called Mourvedre. It was a very successful variety in Algeria but it is not clear how much of the 20 000 ha formerly planted has survived since independence. Australia may well rank next in plantings as there are only about 135 ha in California and very little elsewhere. Most of the 630 ha in Australia are in South Australia, with only 110 ha in New South Wales and 30 ha in Victoria. The name Balzac is used at Corowa, and at Great Western the variety is called Esparte.

Mataro is a vigorous variety with a very upright habit of growth, although the strong tendrils often contort the tops of the shoots. It has a late bud burst and moreover recovers well after frost. The leaves are medium, dark green, entire to 3-lobed with rather coarse teeth, and tufted hairs on the lower surface. Bunches are medium, conical and compact, often with a woody stalk. Berries are small to medium, round with a heavy bloom and thick tough skin.

Mataro alone gives wines which are rather neutral in flavour and can be very astringent. Possibly for this reason the variety has not been very highly regarded in Australia. However, it combines well with other varieties and is an important component in some European wines. In Bandol, for example, where Mataro was largely replaced by Grenache and Cinsaut because of grafting difficulties during reconstitution, the controlled appellation regulations have recently been amended to provide for an increase from 20% to 50% in the minimum proportion of Mataro grapes to be used.

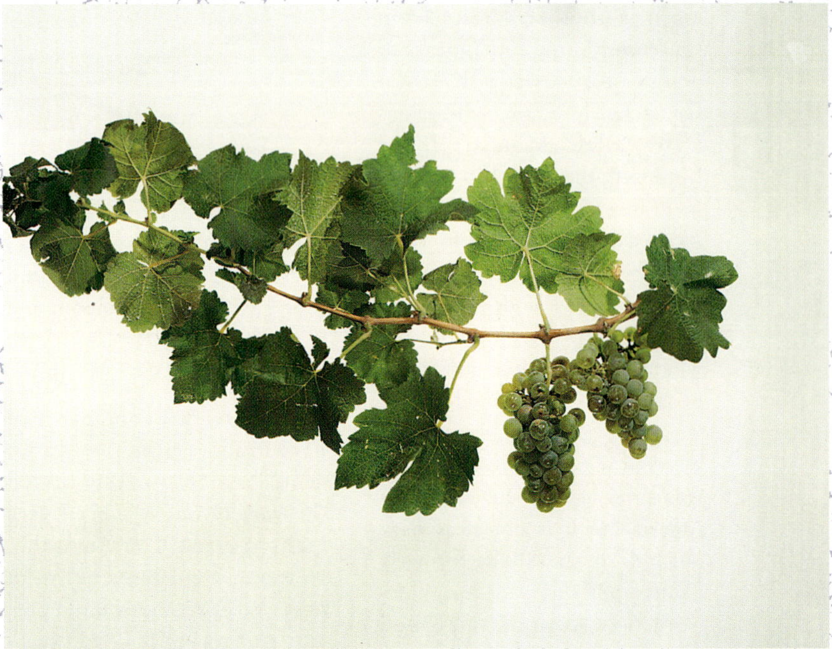

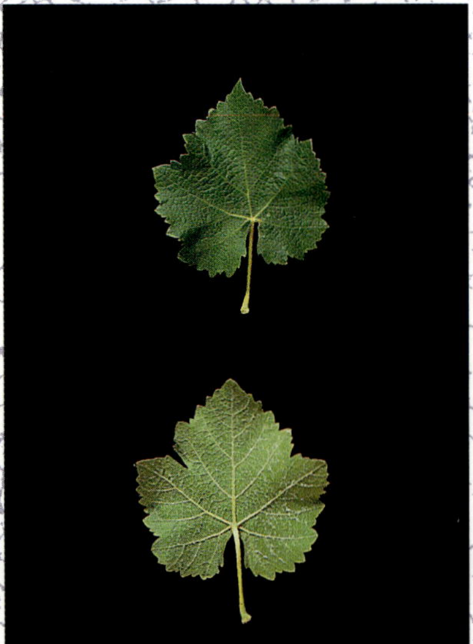

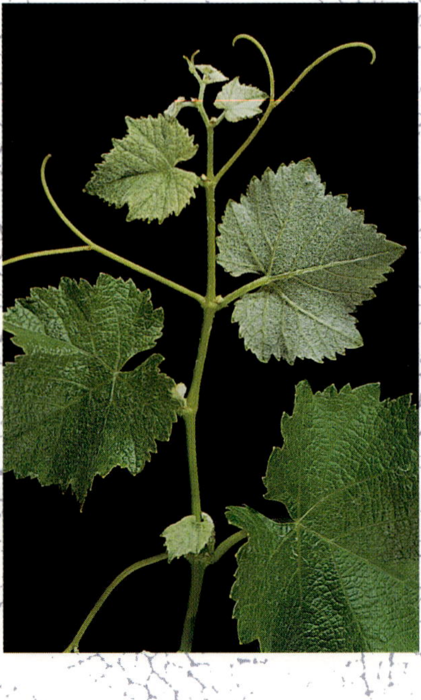

Mauzac

Mauzac is a major white variety in France although confined to a relatively small area in the south west. In 1988 there were 5709 ha of Mauzac in France, compared with, for example, 19 870 ha of Chardonnay, 11 347 ha of Melon and 12 027 ha of Sauvignon Blanc. Mauzac is used in the wines of controlled appellation of Gaillac and Limoux. Mauzac has not attracted the same attention outside France as the other varieties. It has probably been present in Australia for 150 years but only small experimental plantings have ever been made.

Mauzac is a fairly vigorous variety with an upright habit of growth. Leaves are small to medium, almost entire with a characteristic triangular terminal lobe. They are thick, somewhat rough and undulating, with tufted hairs on the lower surface. The petiolar sinus is closed, often with overlapping edges.

Bunches are small to medium, compact, sometimes winged, with a short, often woody stalk. Berries are medium and round with a tough skin. Mauzac has the reputation in France of being relatively resistant to powdery mildew and black spot, and not a difficult variety on which to control downy mildew.

Mauzac is used for still wines, both dry and sweet, and for sparkling wines. Fresh young wines from fruit that is not too mature have an apple-like character reminiscent of cider. Wines from fully mature grapes tend to oxidise rather readily.

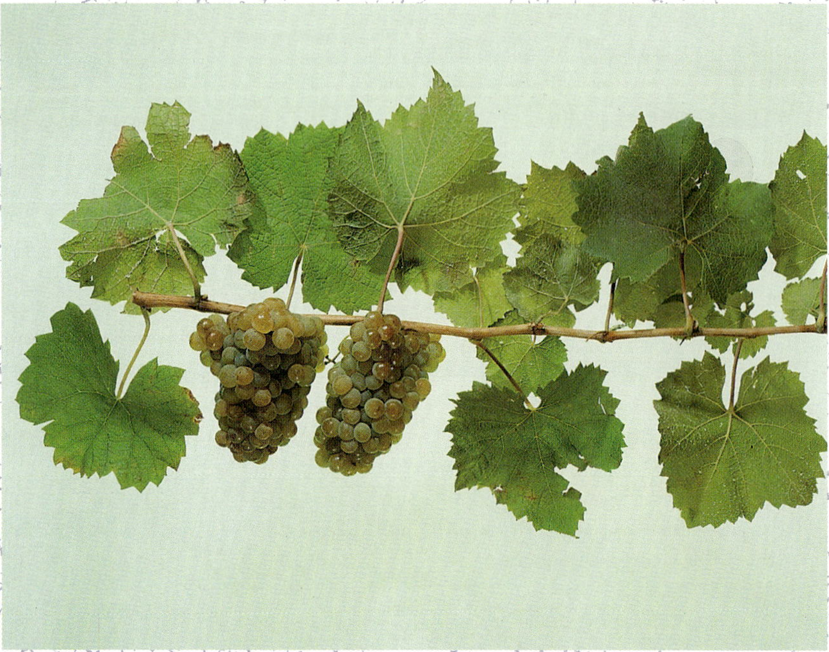

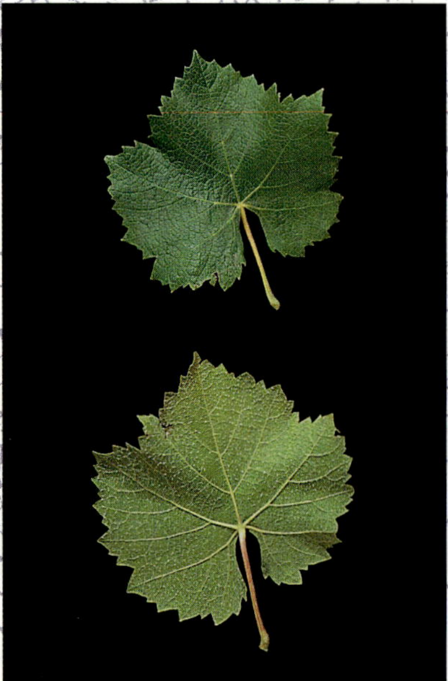

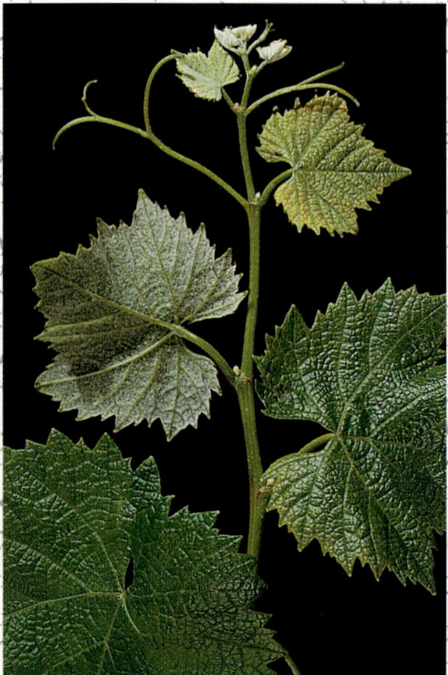

Melon

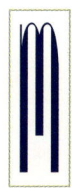

Melon is the official French name for this variety and is used in Burgundy where it originated. However, very little is now grown in this area, compared with about 11 000 ha in the lower valley of the Loire, where it is known as Muscadet. In California, where it is erroneously called Pinot Blanc, there were about 650 ha in 1992. It was imported into Australia from California in 1962 under the name of Pinot Blanc so care is needed to ensure that it is not confused with the true Pinot Blanc which has been imported from Europe more recently.

Melon is a variety of only moderate vigour and yield. The buds burst early in spring and the fruit ripens early and is susceptible to bunch rot. The leaves are medium, entire to slightly 3-lobed, rough, with a characteristic rolling back at the edges and a few tufts of hair on the lower surface. The petiolar sinus is a narrow lyre. The bunches are small and compact with short stalks and small, round berries with rather thick skins.

The wines of Muscadet produced in the Loire Valley are dry and fresh with a good bouquet, and are highly regarded as wines to be drunk young. Good wines have also been produced in California, where a high content of tannin in the skins has been noted and special care in wine-making has been recommended to prevent darkening.

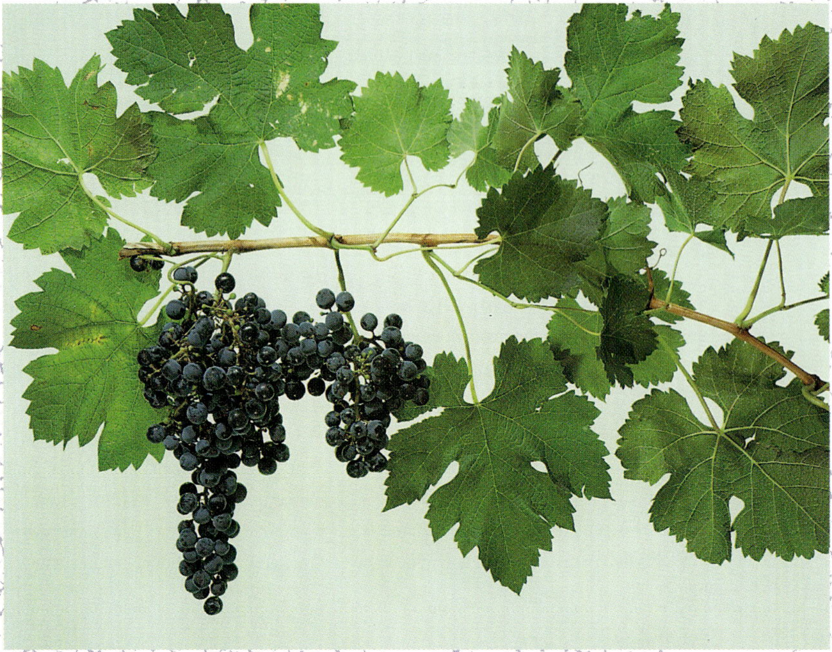

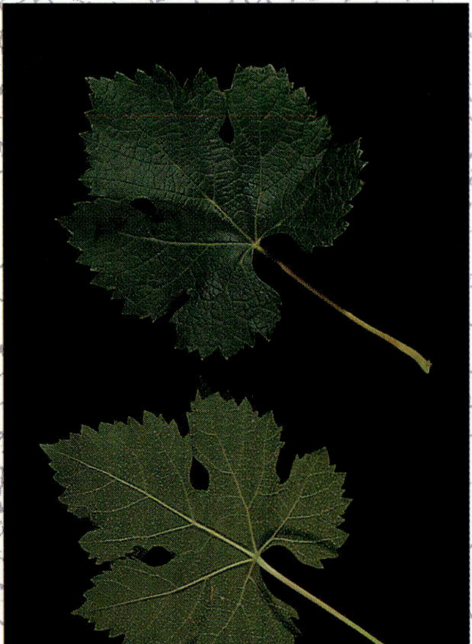

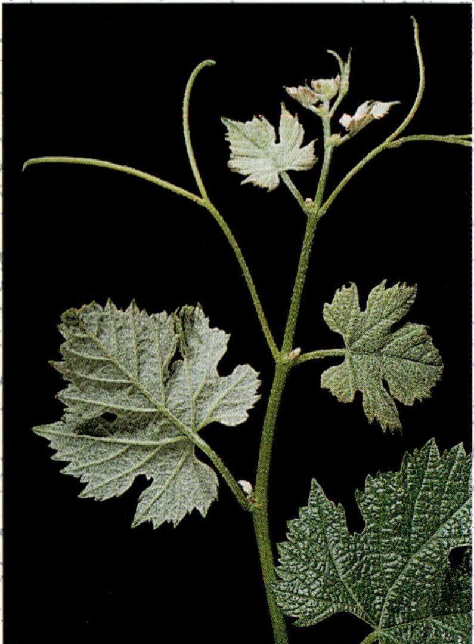

Merlot

Merlot is the principal black variety of the Bordeaux area and is now also recommended in the south of France. The area planted in France increased from 16 800 ha in 1958 to 60 000 ha in 1988. The variety has been introduced into other European countries and is tending to replace local varieties in eastern Europe. In Italy it is now recommended in more provinces than any other single variety although it is far from surpassing Italian varieties such as Barbera and Sangiovese in area or production. In California, from almost no plantings before 1970, the area of Merlot increased to 870 ha by 1982 and to 4050 ha by 1992, of which 1400 ha were still not bearing. It is a minor variety in Chile and Argentina. So far no early introduction into Australia has been traced. There were 739 ha of Merlot planted in Australia by 1993, of which 127 ha were not yet bearing.

Merlot is a vigorous variety although it seems to be one of the most sensitive to salinity. It has medium, 3- to 5-lobed leaves often with a tooth at the base of the lateral sinuses, with scattered tufts of hair on the lower surface and a U-shaped petiolar sinus. Bunches are medium, cylindrical, sometimes winged, and loose with small, round berries. Production is usually good, but under unfavourable conditions at flowering, setting may be very poor.

The wine of Merlot has a distinctive character clearly related to that of the Cabernets. It has good colour but it is softer and ages more quickly than Cabernet wines. Although it may not be used alone, it may be blended with the Cabernets in the finest of the wines of controlled appellation of the Bordeaux region.

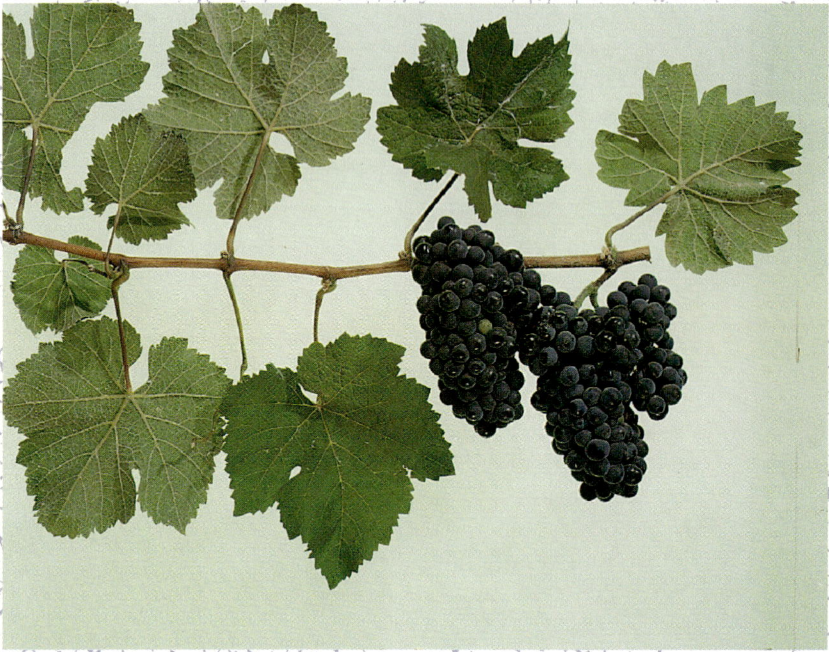

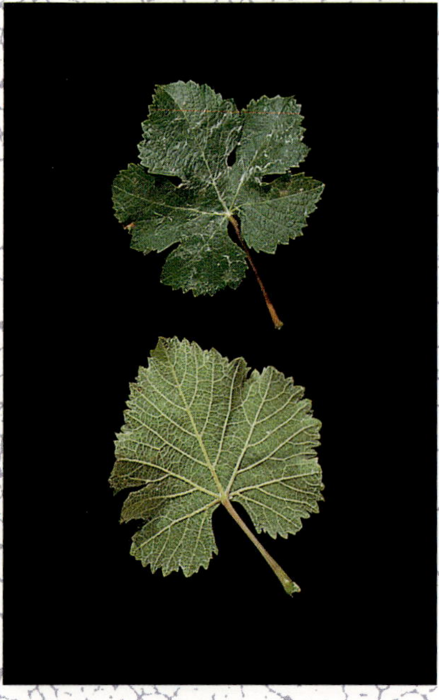

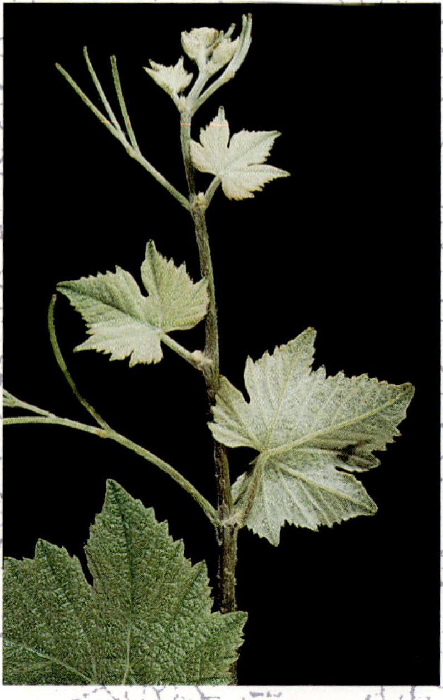

Meunier

Meunier, or Pinot Meunier as it is often known, appears to be a simple sport of Pinot Noir and could have been established independently on a number of occasions. Plants from such sports may be chimaeras (e.g. Meunier in the outer layers and Pinot Noir internally) and when propagated may give plants of both varieties as well as further chimaeras, suggesting instability in the variety. The greatest area of the variety is in France. As with Pinot Noir, the area has been expanding from 5600 ha in 1958 to 11 100 ha in 1988, nearly all in the Champagne area where it is the major variety. In Germany there are about 2000 ha, mostly in Wurtemburg, where it is known as Schwarzriesling. In Australia, where the name has sometimes been translated to Miller's Burgundy, it is confined to a few small plantings in Victoria. In 1960 there were 22 ha in New Zealand, but it has since lost favour and very little remains.

The most distinctive feature of Meunier, to which it owes its name, is the copious white hairiness on the shoot tips which gives the vine the appearance of having been dusted with flour. Like Pinot Noir, Meunier is a variety of only moderate vigour, taking some years to develop vines of a reasonable size.

Leaves are medium in size, more or less deeply 5-lobed, with a few cobwebby hairs on the upper surface and a complete cover below of short, erect hairs between the veins and longer, recumbent hairs along the veins. Bunches are compact and rather small, but often winged so that they may be nearly as wide as they are long. Berries are small and round with a heavy bloom.

In France Meunier is used for champagne. In Australia it is used to make attractive, light, dry red wines.

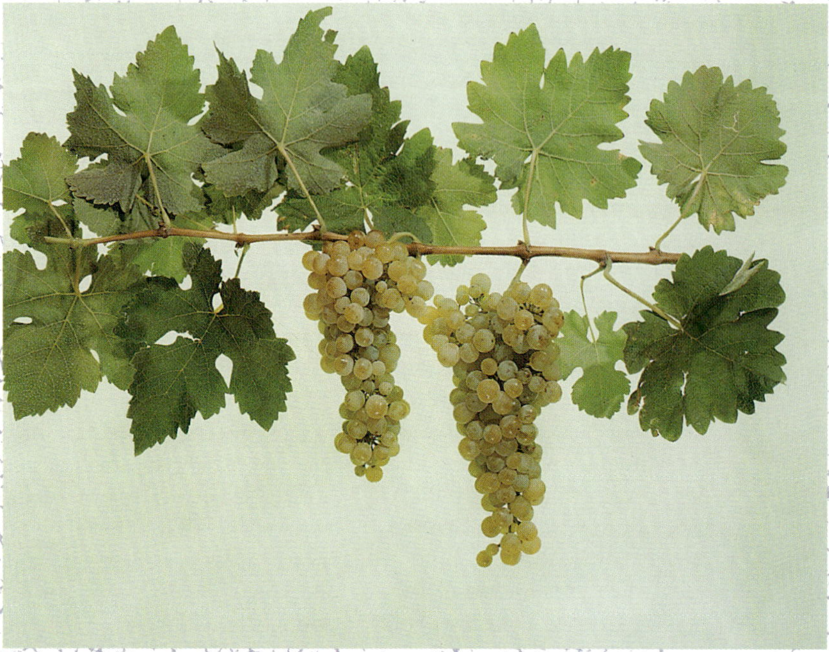

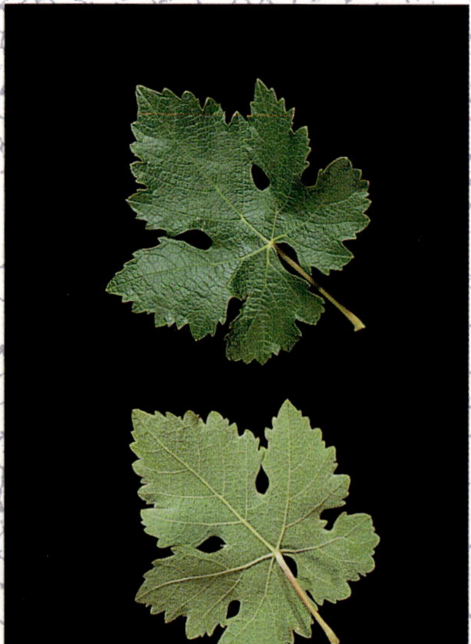

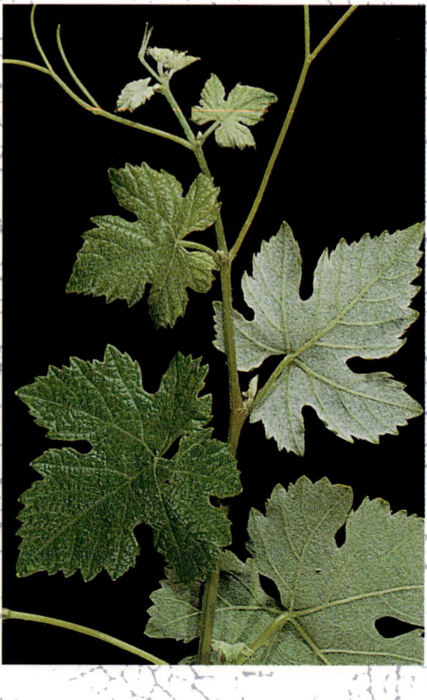

Monbadon

Monbadon comes from France and is found mainly in the Bordeaux and Cognac areas with a little in Provence. It is another variety no longer recommended or authorised, and has declined in area from 1731 ha in 1958 to 129 ha in 1979. It is a minor variety in California, where it is known as Burger, but is maintaining its area at about 900 ha. In Australia there is a little Monbadon in the Corowa–Wahgunyah area.

Monbadon is a variety of rather low vigour with a spreading habit of growth. The leaves are rather small, pale green, 3- to 5-lobed with deep, open lateral sinuses but the lobes of the petiolar sinus overlap at the top. They have tufted hairs on the lower surface. Bunches are large, irregularly conical and compact with medium, round, very juicy berries. In California the fruit is reported to juice very badly during mechanical harvesting, but the large bunches and relatively light foliage make Monbadon an easy variety to harvest by hand.

In France and California Monbadon is used only as an accessory variety, giving light, neutral wines which can be blended with wines from more distinctive varieties. At Corowa–Wahgunyah it accumulates enough sugar to be useful for dessert wines.

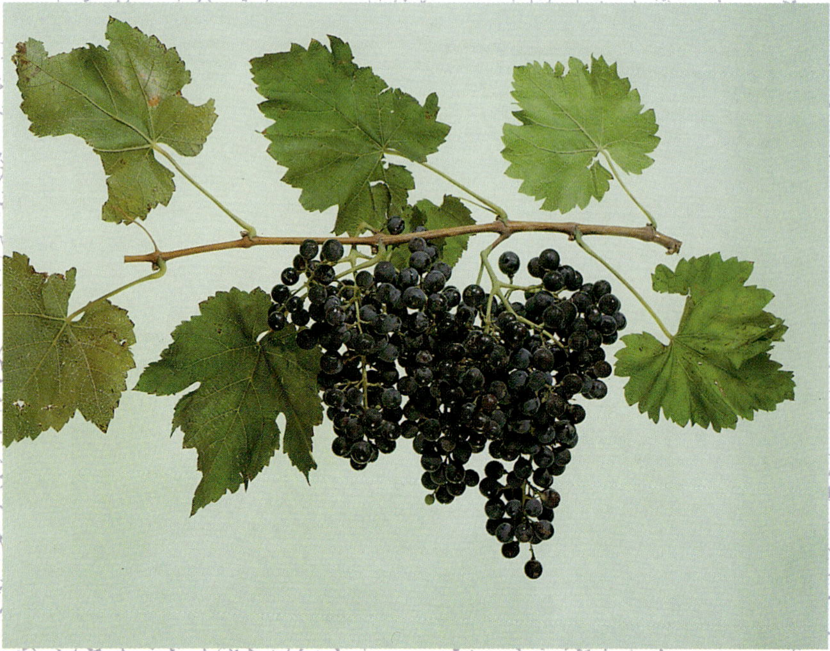
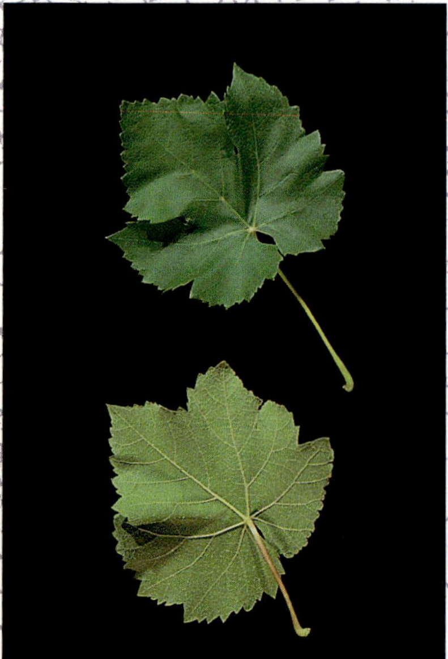
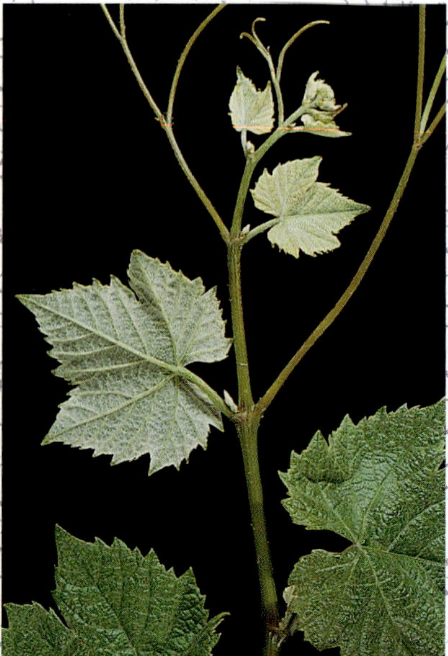

Mondeuse

Mondeuse is a minor variety in the east of France. The area has been decreasing slowly from 1066 ha in 1958 to 173 ha in 1988. In California there has been a more rapid decrease to about 40 ha in 1980. Although the variety is called Refosco in California it is different from any of the varieties known as Refosco in Italy. Mondeuse is grown commercially in Australia on 12 ha in north-east Victoria.

Mondeuse is a variety of moderate vigour which can yield very well, perhaps comparably with Grenache. It has medium, slightly 3-lobed leaves which are undulating and slightly rough and have tufted hairs on the lower surface. In windy weather in spring the lower leaves on the shoots become rather tattered quite early in the season. The bunches are large, often with a long stalk, and can range from rather loose to compact. The berries are small, round to slightly oval, intensely black with a bluish bloom. The fruit tends to be exposed when ripe but it ripens late and does not appear to be adversely affected. It retains a fair acidity even in warm climates. Mondeuse has a better than average resistance to fungus diseases and to rain damage at harvest.

The wines of Mondeuse are notable for their colour and tannin. In California they are used for blending, and the limited quantity of Mondeuse produced in Australia is combined with other varieties. In France vineyards on the slopes with yields of 6 to 9 tonnes per hectare are said to give excellent wines, but those on more fertile soils on the plains where yields surpass 15 tonnes per hectare give wines which are only ordinary.

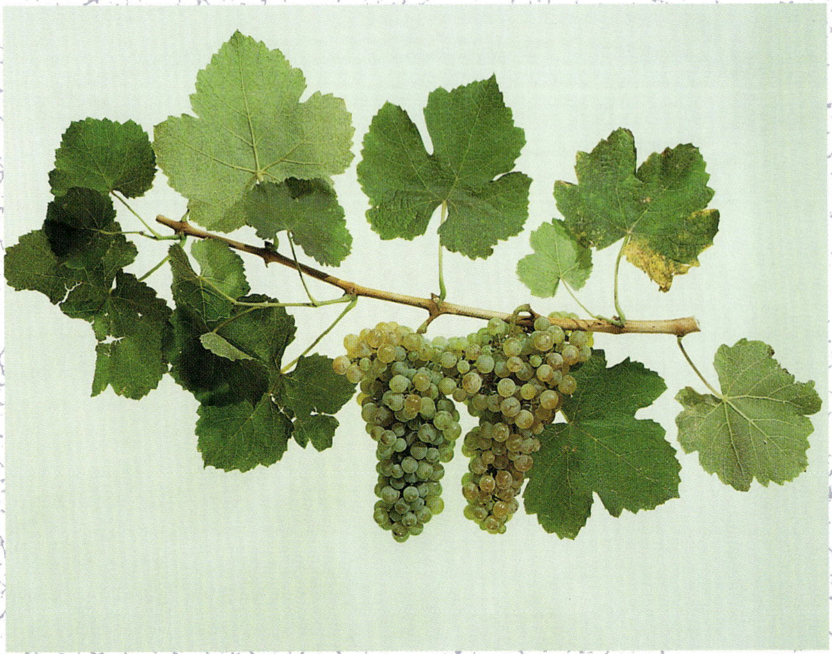

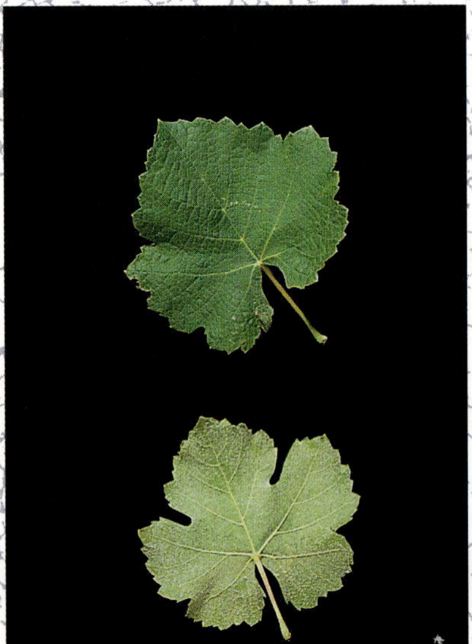

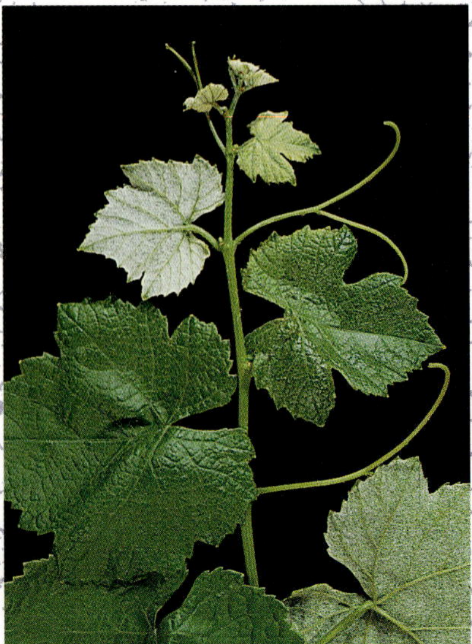

Montils

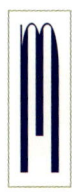

Montils is a minor variety of the Cognac area of France grown mainly close to the coast. Although it remains an authorised variety for the area, plantings have declined considerably. In Australia there are small plantings in the Hunter Valley, some as Montils and some under the name of Aucarot or Aucerot. This name is a corruption of Amarot, a variety introduced by James Busby in 1832, but, as happened in other cases, the plants were confused; Amarot is a large black grape. Montils was also tried experimentally in north east Victoria but the Aucerot in this area appears to be a different, as yet unidentified, variety.

Montils is a fairly vigorous variety with a spreading habit of growth. Leaves are medium, entire to 3-lobed, tending to fold about the midrib, dull green, smooth to slightly rough, with tufted hairs on the lower surface. Bunches are small to medium and well filled, with rather small, round berries.

In France, Montils is considered to give a wine comparable with that of Colombard. At Merbein, Montils usually ripens about 2 weeks later than Colombard and has an excellent low pH. Yields of the two varieties are comparable. Wines from Montils over 4 seasons scored at least as well as those from Colombard, and Montils was one of the most promising varieties in the trials in north-east Victoria.

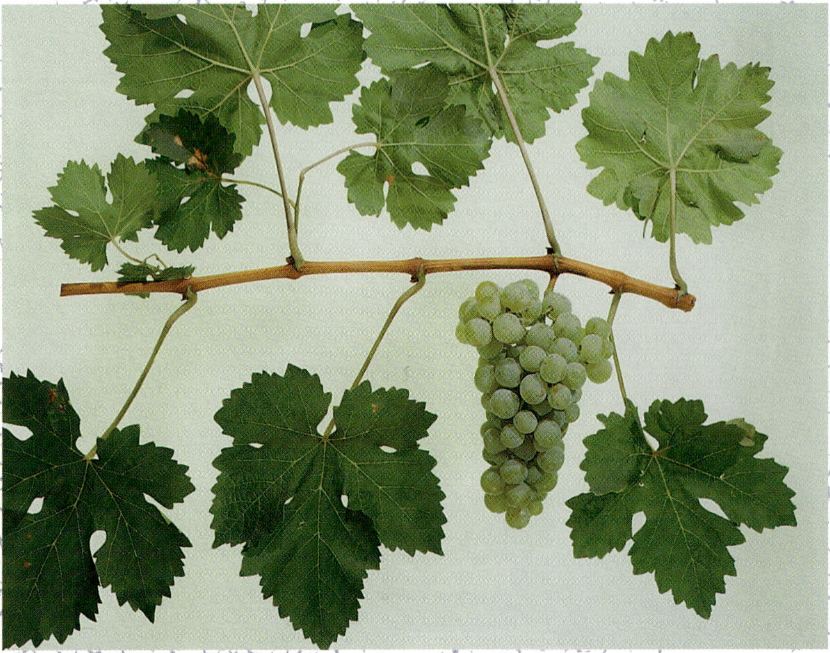

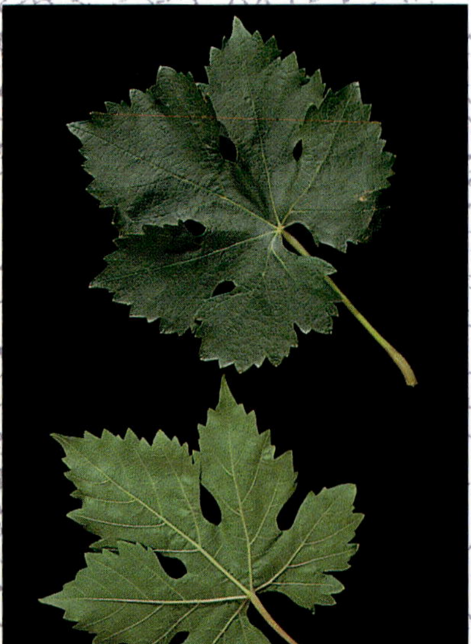

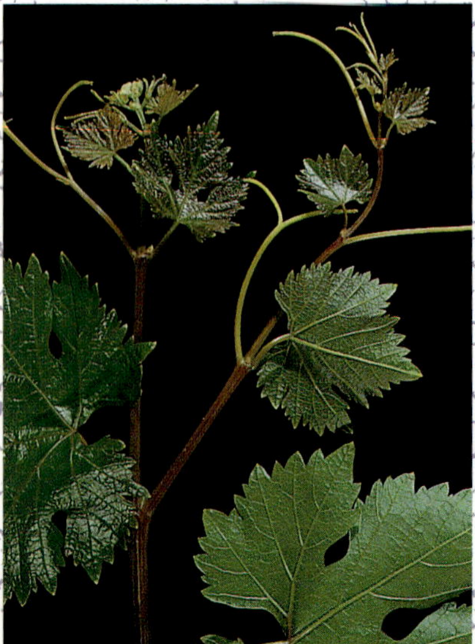

Moschata Paradisa

Moschata Paradisa is the name by which this variety is known in Australia and so far it has not been traced to any variety grown or described overseas. It is known to have been planted commercially only at Mudgee although there may be odd vines elsewhere. The early maturity and rapidity with which the fruit then deteriorates, combined with the softness and slightly unusual flavour of the berries, suggest that there may be a little *Vitis labrusca* in the ancestry of the variety. On the other hand the leaves show no sign of any species other than *V. vinifera*. There is some resemblance to the Malvasia Bianca imported from California but the two varieties are distinct.

Moschata Paradisa is a vigorous variety with a spreading habit of growth and a tendency to produce forked shoots, often at the bunch. Leaves are medium, deeply 5-lobed, folding strongly inwards, with short, bristly hairs mainly along the veins on the lower surface. The petiolar sinus is an open lyre with the sides appearing to overlap because of the folding of the leaf. Bunches are medium, conical and compact with a short stalk. Berries tend to be oblate and show signs of lobing, and are soft with a muscat flavour.

While care is needed to harvest the fruit as soon as it is ripe, Moschata Paradisa can be used to make a pleasant varietal wine.

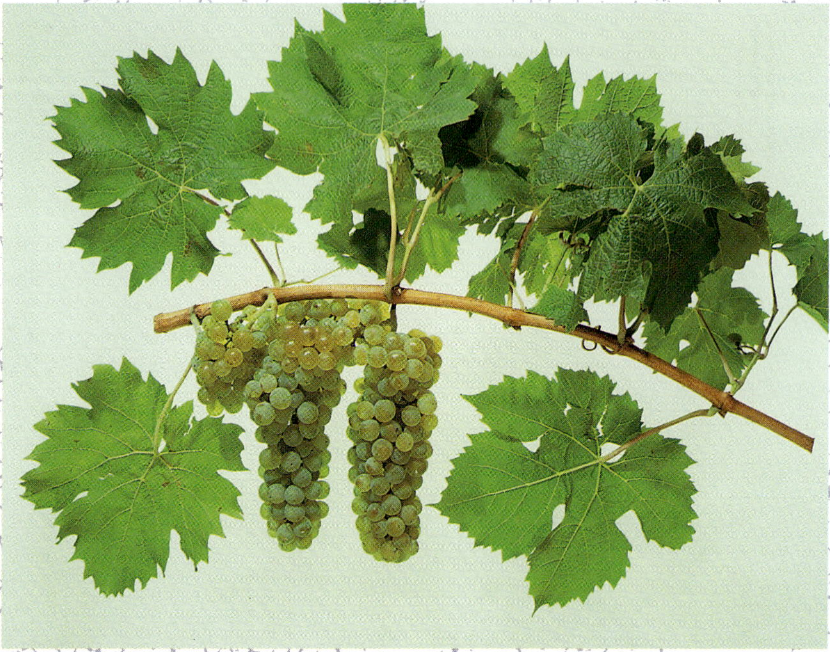

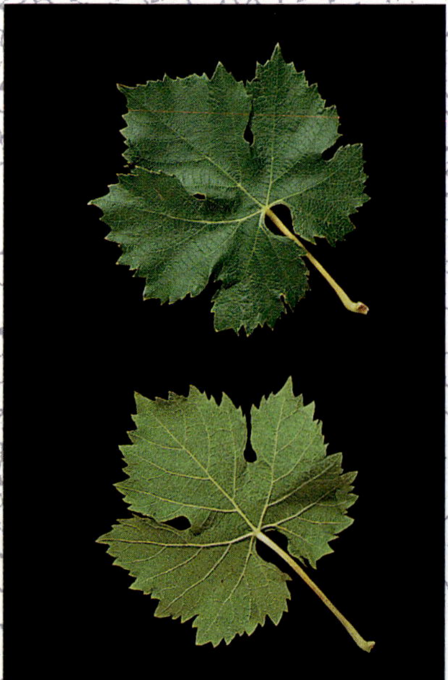

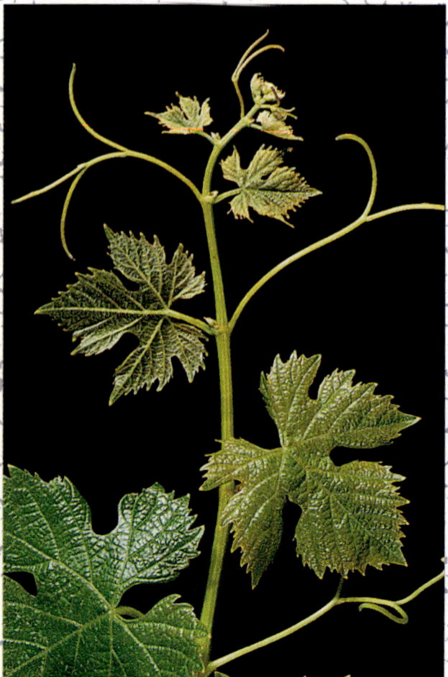

Müller-Thurgau

First introduced into commerce in about 1920, Müller-Thurgau had by 1970 passed Sylvaner and Riesling to become the leading wine grape of Germany. The area has now stabilised at about 25 000 ha. It is also grown in central Europe with about 5300 ha in Czechoslovakia, 4400 ha in Austria and 8000 ha in Hungary, but, apart from a very small area in Alsace, not in France or in North or South America. It comes from a cross made at Geisenheim in 1882 by Dr Müller, a Swiss from Thurgau, who later returned to Switzerland taking his promising seedlings with him. In 1913, by which time its great potential was obvious, it was brought back to Germany for testing. It is supposed to be a cross of Riesling and Sylvaner, but recent DNA typing showed that Sylvaner could not have been a parent of this variety. From its character, some experts think that it is a cross of two Riesling clones, but DNA typing also showed that although Riesling was one parent of Müller-Thurgau, it could not have been both. In Switzerland and some other countries it is called Riesling x Sylvaner in deference to its breeder's wishes. In Luxembourg where there are about 600 ha it is called Rivaner. There has been some very limited planting of the variety in Australia.

Müller-Thurgau is a vigorous and productive variety that matures its wood well. The leaves are medium in size, distinctly 5-lobed and slightly downy on the lower surface. The petiolar sinus is lyre shaped and sometimes cut right in to the veins at the base. Bunches are medium to large, loose to well filled with medium, oval berries. The fruit ripens early and is very susceptible to bunch rot. The vines are more susceptible to winter freezing (at temperatures lower than would occur in Australia) than those of Riesling.

The wine from Müller-Thurgau, although less acid and without the very distinctive bouquet of Riesling, has a definite character and is well regarded among German wines.

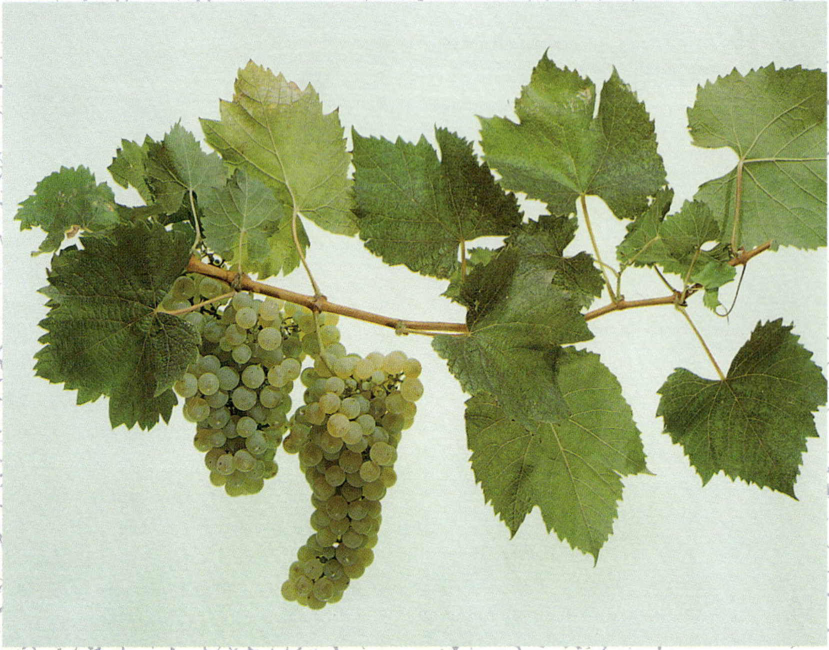

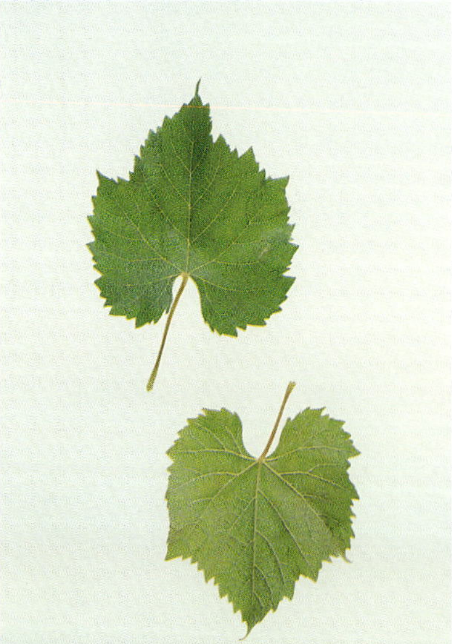

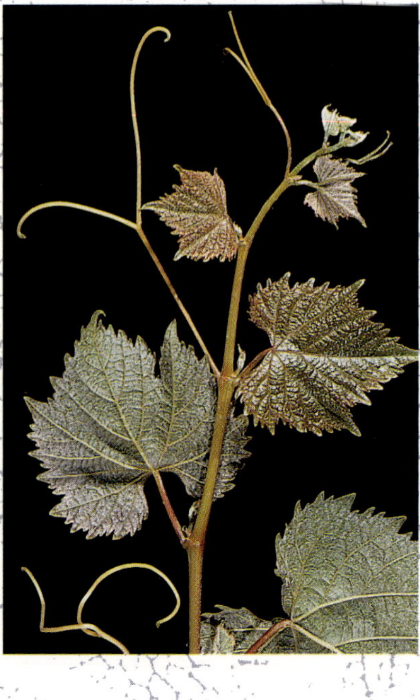

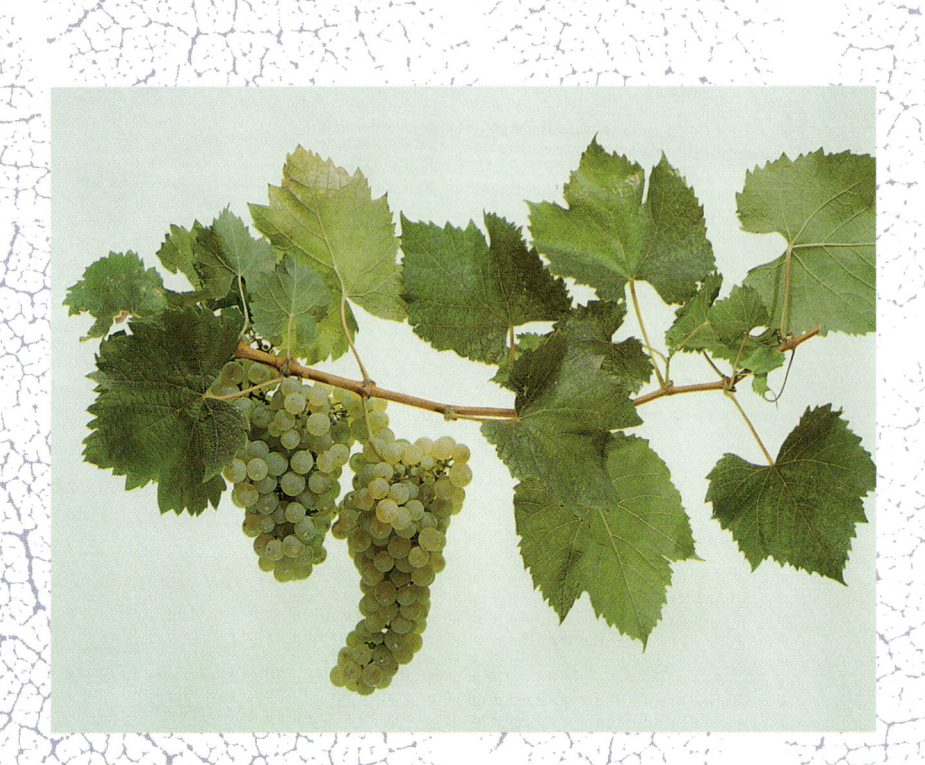

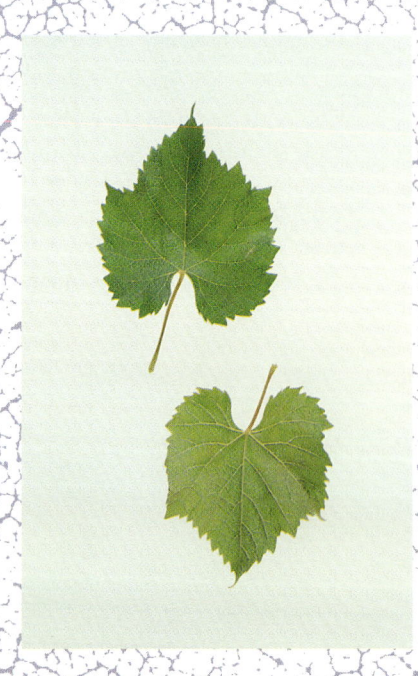

Muscadelle

Muscadelle is a variety of the Bordeaux region where it forms a minor component in the famous wines of Graves, Barsac, Sauternes, etc. There were 5700 ha of Muscadelle in France in 1968 but this decreased to 2700 ha by 1988. Muscadelle has been grown in Australia under the name of Tokay. There were 359 ha of Muscadelle in Australia in 1993. Sauvignon Vert introduced from California has also proved to be Muscadelle. Sauvignon Vert is no longer favoured in California and only about 34 ha remained in 1992. There also appear to be plantings in Hungary, Romania and the Ukraine. Muskadel in South Africa is a different variety, Muscat à petits grains.

Muscadelle is a fairly vigorous and productive variety. The young spring growth has a characteristic translucent reddish appearance. The leaves are large, rough and undulating, 3-lobed with the petiolar sinus sometimes closed, and with a little hair on the lower surface. The bunches are fairly large but often rather loose. The berries are rather small, round and speckled with brown at full maturity. The fruit ripens early and can attain a very high sugar content.

In France, Muscadelle gives a wine with a marked bouquet somewhat reminiscent of muscat and only a small proportion is needed to achieve the desired degree of this character. When used on its own in California, Sauvignon Vert gives a bitterness on the palate which is found objectionable. The use of very ripe, partially raisined grapes for a sweet fortified wine seems to be specific to Australia and is very successful.

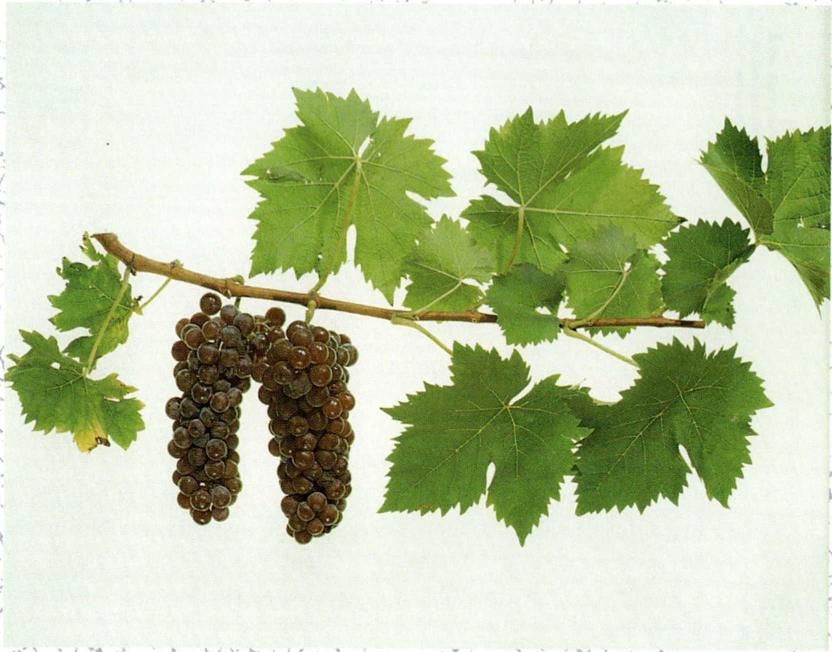

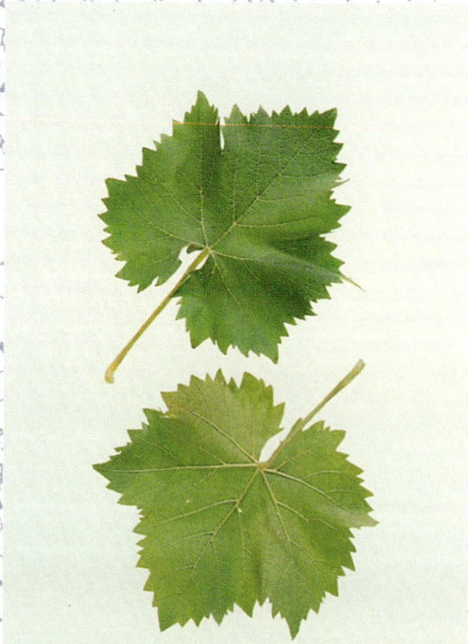

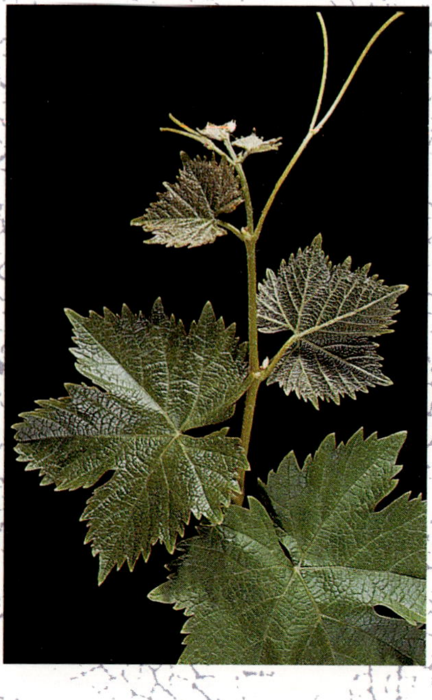

Muscat à petits grains

Muscat à petits grains is the official French name for this variety and means simply Muscat with small berries. The Spanish name Moscatel Menudo has a similar meaning. The names Muscat de Frontignan and Muscat d'Alsace, used in the parts of France concerned, are not acceptable to the Office International de la Vigne et du Vin for use elsewhere because they refer to place names. This also applies to the Italian names Moscato d'Asti and Moscato di Canelli. Rather surprisingly, the corruption Frontignac may be used. There are three colour variants of the variety — white, rose and red. The coloured forms seem to mutate readily from one to the other and to white, but there seem to be two types of white — one which is stable and one which mutates readily to the coloured forms. Possibly chimaeras, in which the genetic make up of the berry skin differs from that of the flesh, are involved.

The white form seems to predominate in Europe, mostly as the stable type. There are 3000 ha in France, probably about the same in Italy where the official name is Moscato Bianco, as well as plantings in other Mediterranean countries. The name Muscat Blanc in California, where there are 485 ha, seems to indicate that only the white form is used there and the 400 ha in Argentina are listed as Moscato d'Asti among the white varieties. In South Africa, on the other hand, most of the 1020 ha of Muskadel, as the variety is known there, are of the red form.

There are 330 ha of Muscat à petits grains in Australia, nearly half of them in South Australia, and nearly all of the rest evenly divided between Victoria and New South Wales. A fair proportion is the red form and the name Brown Muscat used in north-east Victoria is very appropriate.

Muscat à petits grains is moderate in vigour and more or less erect in growth. It has small to medium, 3-lobed leaves with sharp angular teeth and hairs only along the veins on the lower surface. Bunches are medium, more or less cylindrical but broader at the top, well filled to compact. The berries are medium or a little less, round, with a strong muscat flavour, bronzing where exposed to the sun. The white form has yellowish green berries and the rose is intermediate between that and the red, which is illustrated. The fruit ripens early and if left on the vine wilts to give a very high sugar concentration.

Even the red form of Muscat à petits grains does not have enough colour to make a red wine. However, the variety can be used to make a range of excellent wines, from highly flavoured sparkling wines to luscious fortified wines. It can also be used to accentuate the flavour of dry white wines made from other varieties.

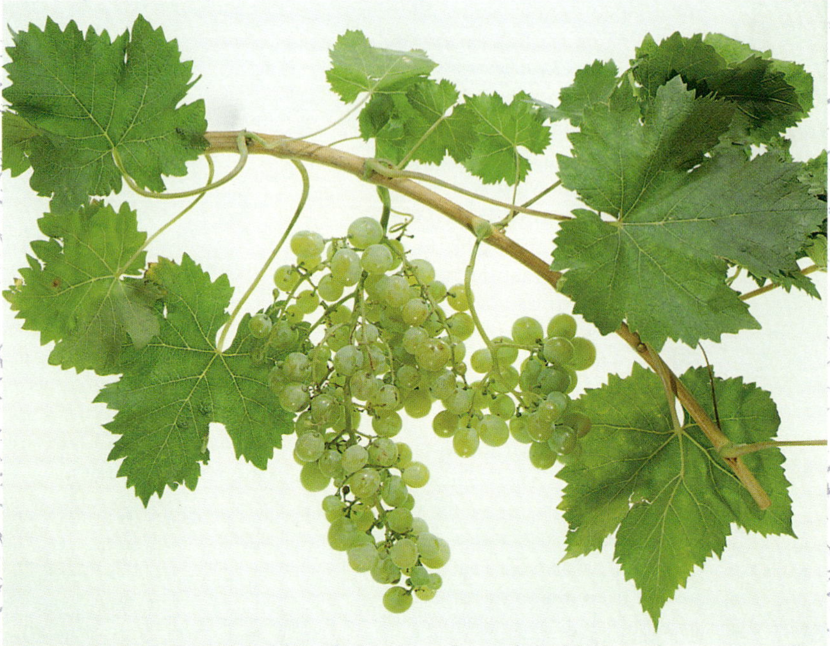

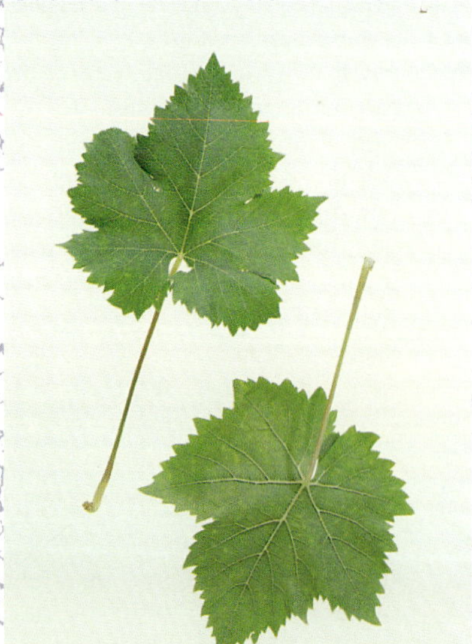

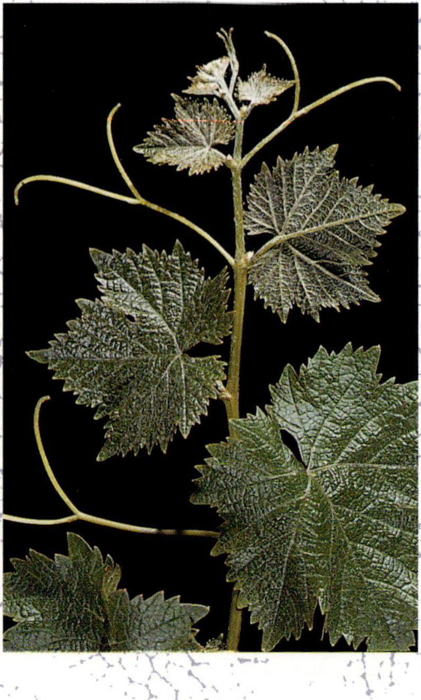

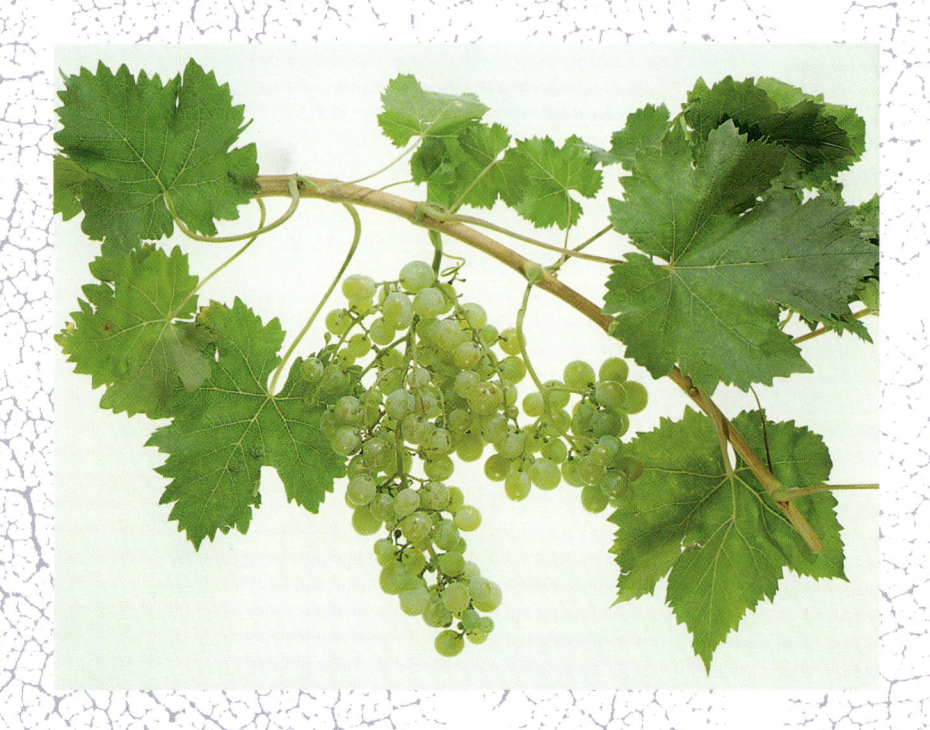

Muscat Gordo Blanco

The name used for this variety in Australia comes from Spain and means simply 'fat white muscat'. The name best known internationally would probably be Muscat of Alexandria. The variety is also called Moscatel de Malaga in Spain, Muscat de Setubal in Portugal, Zibibbo in Italy and Hanepoot in South Africa. Although widely grown it does not cover a very large area in any country. There are about 3400 ha in Australia, 2000 ha in California, 5700 ha in South Africa, 10 000 ha in Argentina, 15 000 ha in Spain and 3000 ha in France.

Muscat Gordo Blanco is not a vigorous variety, particularly once it is in bearing. Unless the permanent framework is fully formed before the first crop, it becomes very difficult to complete it. The vines are best closer planted than those of other varieties, no more than 1.6 m apart in the row. The variety is best pruned to short spurs of one clear bud.

The leaves of Muscat Gordo Blanco are small to medium, mainly 3-lobed and of a rather dull green colour. They are naturally hairless but frequently carry blisters, the inner surface of which is densely hairy, due to the activity of the erinose mite. Bunches are large but often rather loose or even straggly with distinctly uneven berry size. Berries are medium to large with a tough skin bronzed where exposed to the sun, firm flesh, hard seeds and a strong muscat flavour. The fruit can attain a very high sugar content in the warmer areas but is then low in acid and has a high pH.

Muscat Gordo Blanco is a true multipurpose grape. It is used as a table grape, dried for raisins, which may be de-seeded for use in baking and confectionery or kept as clusters for dessert use (muscatels), and crushed for unfermented grape juice as well as for wine. In the Australian wine industry it is used for fortified sweet wines of the type known as cream sherry, and for table wines, often in conjunction with a more neutral variety such as Sultana.

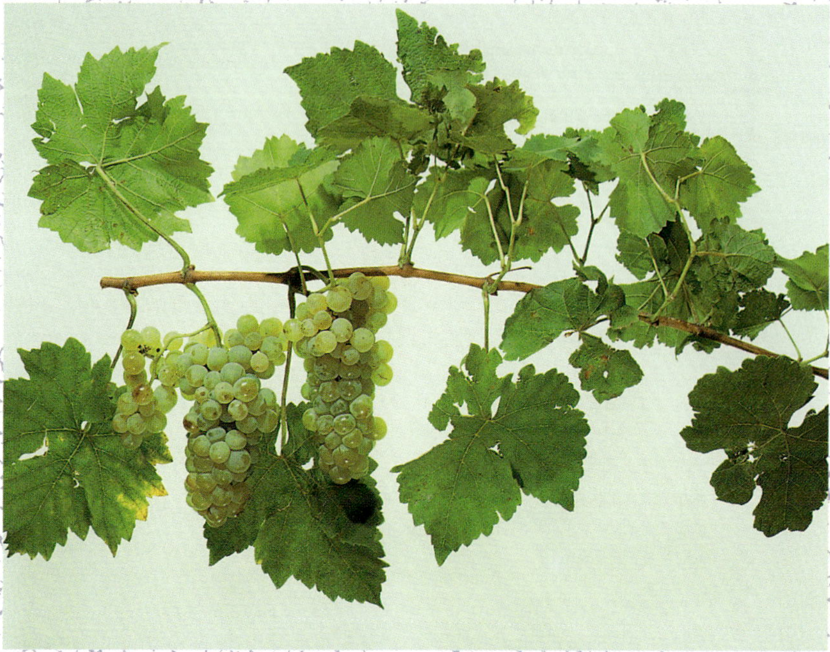

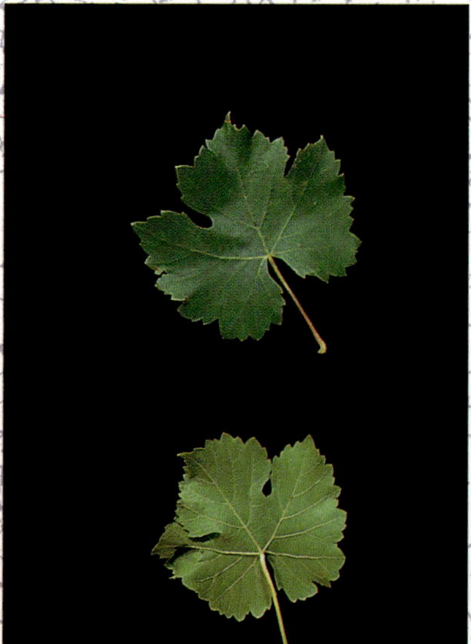

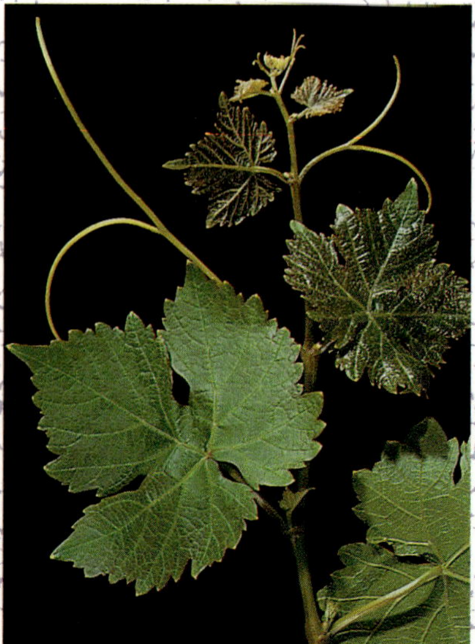

Muscat Ottonel

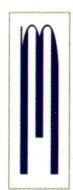

Muscat Ottonel is an early ripening variety grown in Austria, Germany and in Alsace in France. The major plantings are in Austria where there were 1371 ha planted in 1981. There were 429 ha planted in France in 1988, down slightly from the 447 ha in 1979 but considerably more than the 192 ha recorded in 1968. There is very little grown in Australia. Muscat Ottonel was propagated from a seedling by Robert Moreau and its parents are uncertain although probably Chasselas x Muscat de Saumur.

In Australia both yields and sugar levels have been low. However, the wines have an attractive, delicate muscat flavour and have been rated highly by taste panels.

Opening buds are cottony, reddish and full. The young leaves are glabrous, shiny and very red. Leaves are small, orbicular and puckered. The petiolar sinus is a narrow lyre shape with the lobes sometimes overlapping. The teeth are pointed and medium in size. The veins on the lower surface are lightly hairy and uneven. The shoots are glabrous and violet tinged where exposed to the sun, the tendrils are very long like those of Chasselas. Bunches are small and loose. The berries are round, medium in size, yellow in colour when ripe with a subtle muscat flavour.

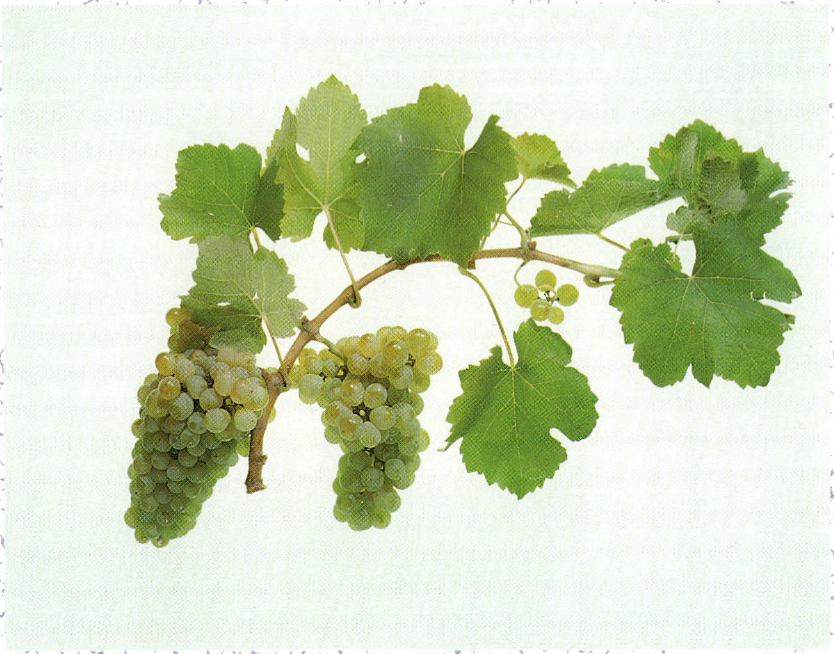

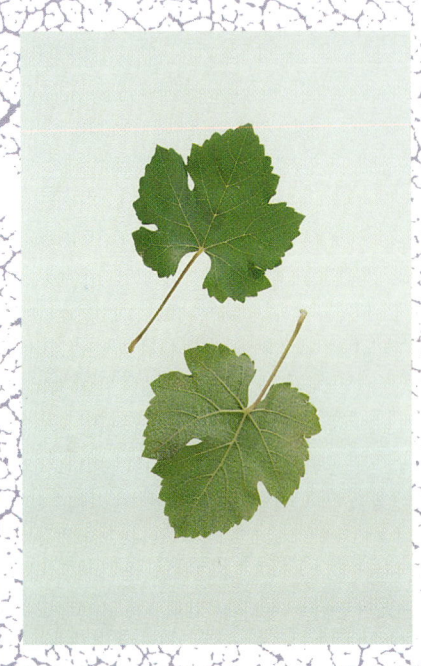

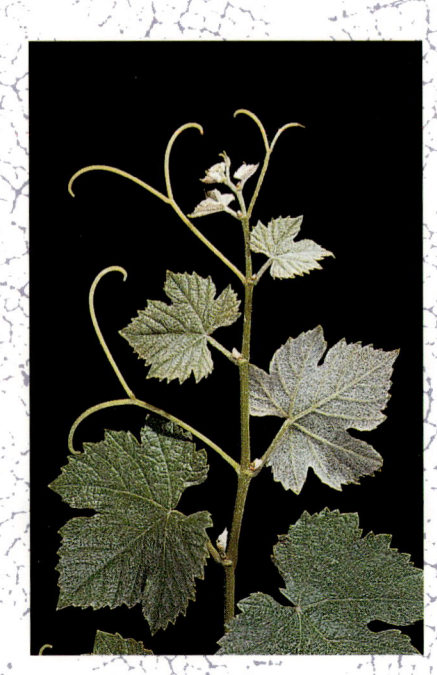

Ondenc

Ondenc is a rather obscure French variety which has been defined only in the last 40 years. It has been grown in different areas of the south-west of France under different names which were not recognised as belonging to the same variety. The name Ondenc comes from the area near Toulouse, where only about 30 ha of the variety remain. In Armagnac it is known as Piquepout de Moissac; before about 1940 it was apparently not distinguished from Folle Blanche, which is also known as Piquepout in this area. There is more in the Bordeaux area, mostly near Blaye, Fronsac and Bergerac, where the name Blanquette seems to have been used. A selection from Fronsac was taken to the Cognac area about 1820 as Blanc Selection Carrière. It was imported into Portugal and California under this name but was not widely planted. In the Pyrenees the names Dourec and Plant de Gaillac were used but very little now remains.

In Australia it has been known as Sercial in South Australia and as Irvine's White in Victoria. There were 23 ha of Ondenc in Australia in 1990. It was probably among the many varieties called Piquepoule collected by James Busby in 1832 and the Victorian plantings may come from this source. The identity was lost and the name Irvine's White commemorates the vigneron at Great Western who made the first substantial plantings. It appears in the Rutherglen collection as Blanc Select, so was presumably imported at some time as Blanc Selection Carrière. The confusion in South Australia could have arisen if Blanc Selection Carrière and the true Sercial, which is a different variety grown mostly on the island of Madeira, had been imported from Portugal together.

Ondenc is a variety of only moderate vigour and productivity. It has dark green, rather rough leaves with tufted hairs on the lower surface. They are generally 3-lobed, but very characteristically some leaves are unsymmetrical and one half can be almost entire while the other is deeply lobed. Bunches are small to medium, well filled, generally with a woody stalk. Berries are medium, short oval, yellowish green to gold with dark blotches, soft and juicy but with a tough skin.

In Armagnac Ondenc is used for brandy, elsewhere in France for white table wines. It is regarded as too neutral to be used on its own and is generally combined with varieties of more character such as Sauvignon Blanc. Its use in sparkling wine seems to be a very successful Australian innovation.

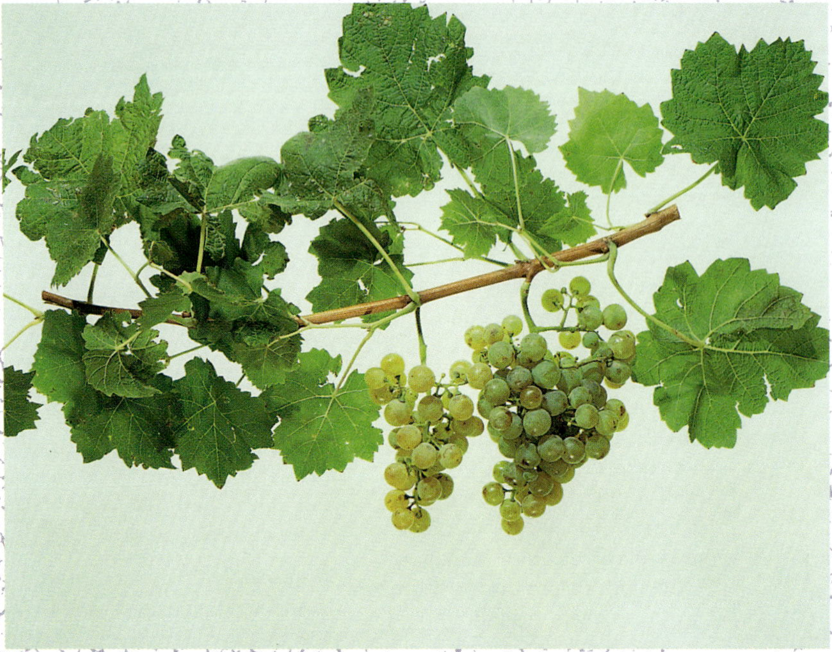

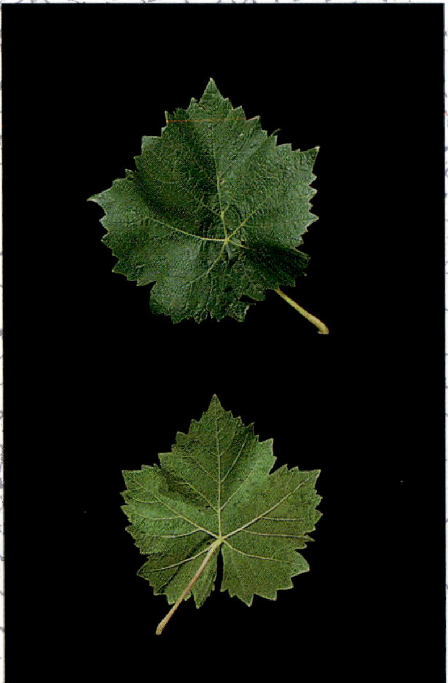

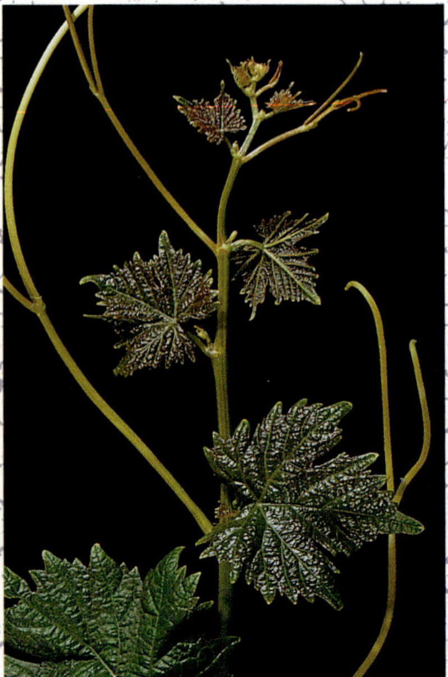

Orange Muscat

Orange Muscat is an old variety which has been used as a table grape in Europe. Its French synonym, Muscat Fleur d'Oranger, translates to Orange Blossom Muscat derived no doubt from the subtle aroma of its juice. Other synonyms are Muscat Primavis and Muscat de Jesus. It has not been widely planted in the world. There were 25 ha in California in 1992 and probably about the same in Australia. One source suggests that it is Syrian in origin.

Orange Muscat is a variety with medium vigour. The opening buds are cobwebby but the young leaves on the growing tip are glabrous and deeply bronzed. The leaves are medium in size, orbicular, entire to 5-lobed, dark green, puckered and bullate. The petiolar sinus is closed and overlapping. The teeth are pointed and large. Green shoots are glabrous with coloured nodes and striped on the back. The mature canes are brown with stripes and mauve coloured nodes. The bunches are medium in size and compact with white, spherical, medium sized berries.

The variety is sensitive to oidium and the berries split easily. It has been used commercially in Australia to produce high quality white wines with distinctive flavour.

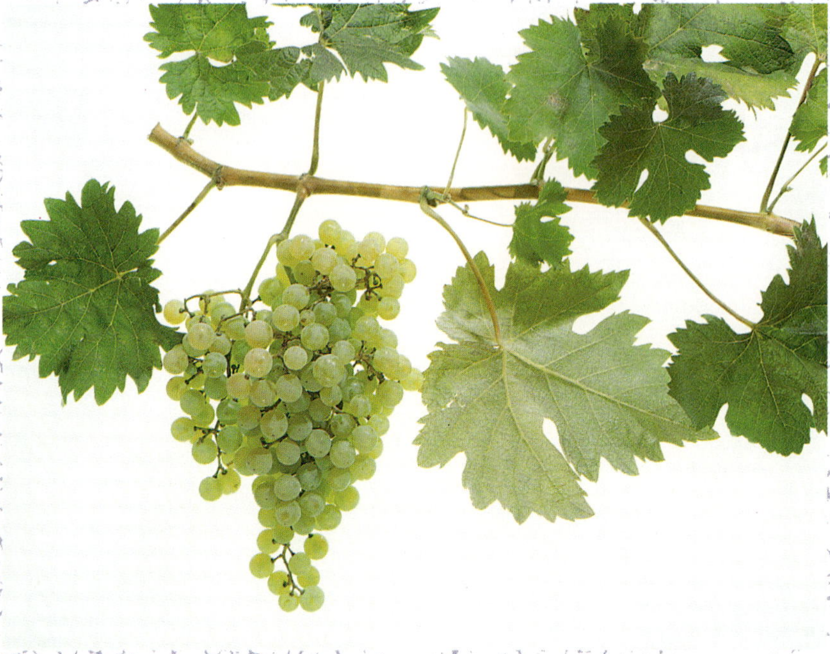

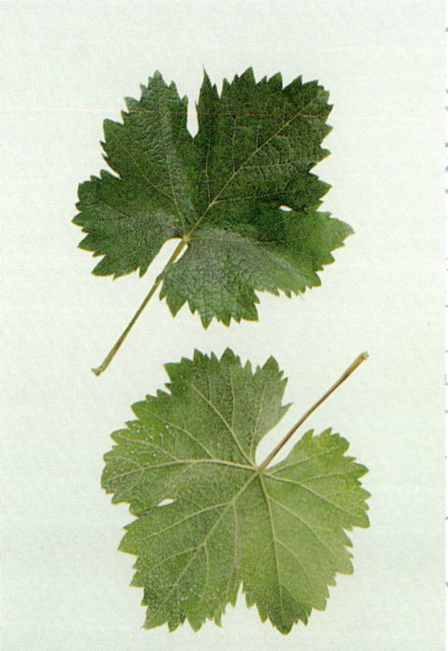

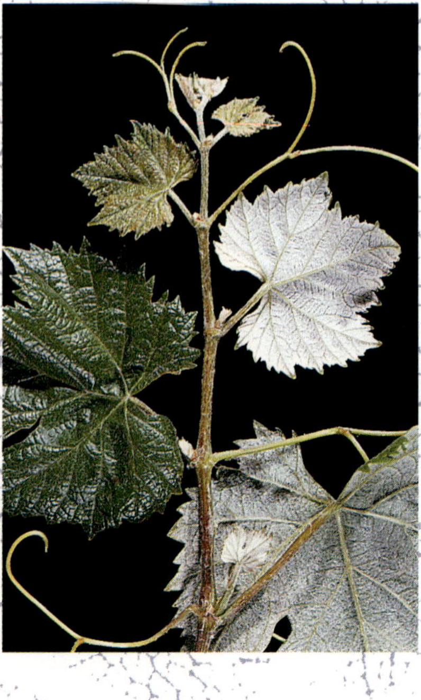

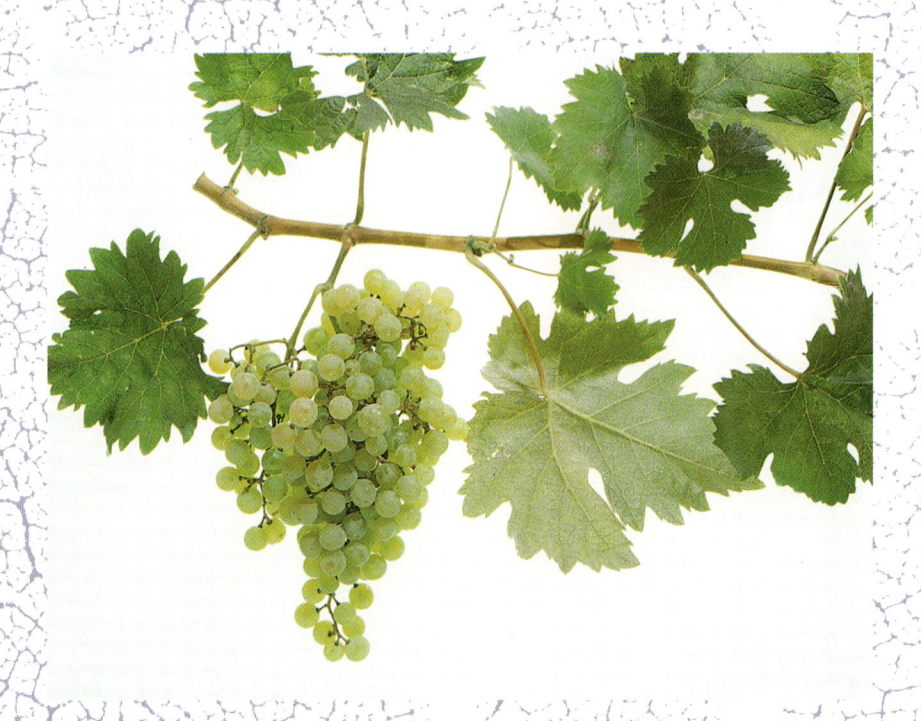

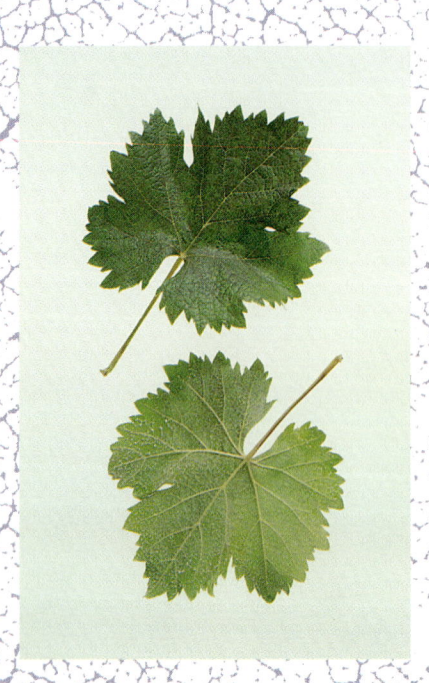

Palomino

Palomino is an important Spanish variety, providing about 90% of the grapes used for sherry. There are about 57 000 ha of Palomino in Spain of which about three quarters are in the sherry area. In South Africa, where it was formerly known as White French, it covers the fifth largest area of any variety, about 5200 ha. Elsewhere the areas are rather small: 470 ha in California where it has sometimes been erroneously called Golden Chasselas; 771 ha in France where it is known as Listan; and 700 ha in Australia where it has sometimes been known as Paulo. It should not be confused with the so-called Common Palomino in Australia, which is in fact Cañocazo.

Palomino is a fairly vigorous and rather upright variety. It has medium to large, distinctly 5-lobed leaves with prominent angular teeth. The leaves are dark green above and densely hairy below. The leaf stalks are generally flushed with red and this colour sometimes continues into the base of the veins. The bunches are large, loose to well filled, with the stalk often flushed with colour like the leaf stalk. The berries are medium, round, firm with a tough skin and golden in colour when ripe. The variety yields well and at ripeness has a high sugar content but low acidity and neutral flavour.

Palomino is better suited to the production of fortified wines than of table wines and is a preferred variety for premium dry sherries in Australia.

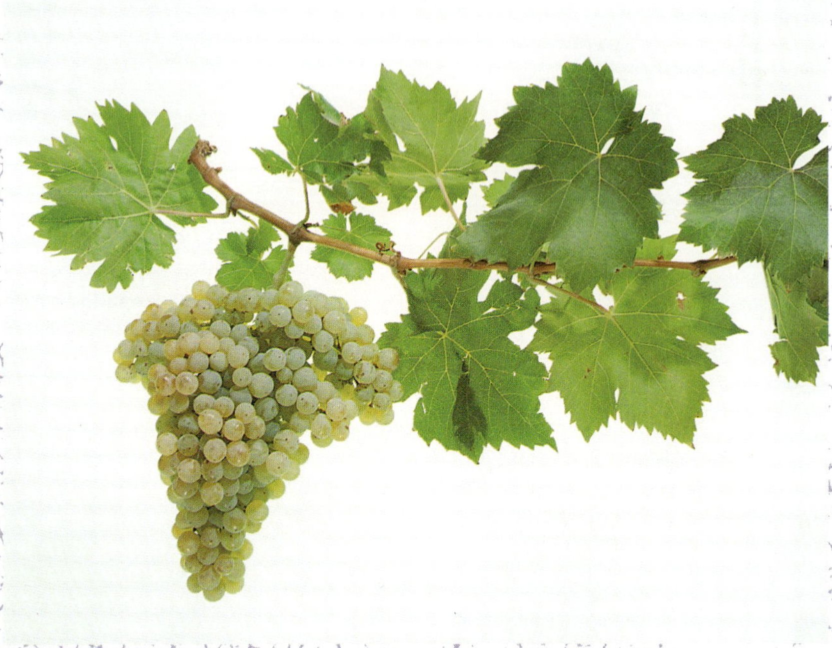

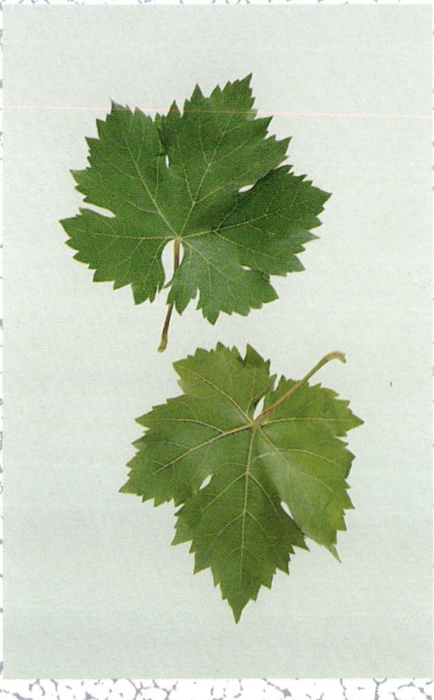

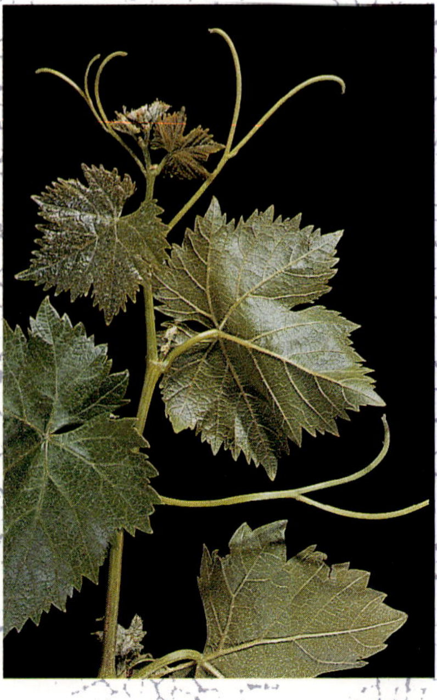

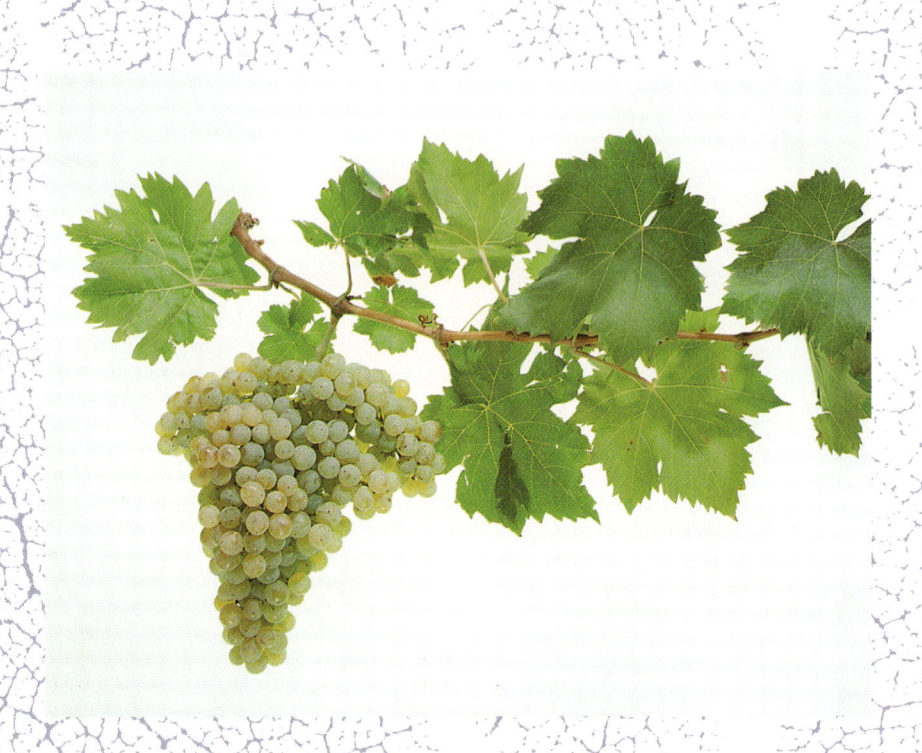

Pedro Ximenez

Pedro Ximenez also comes from Spain, where there are plantings of about 33 000 ha mostly in the Estremadura, Andalusia and Levant regions. While it is a permitted variety for sherry, only a small proportion of the plantings are in the sherry area. It forms the most important recognised white wine variety in Argentina, where there are 24 000 ha. It is declining in California, where there were less than 100 ha by 1976 and no new plantings had been made for many years. A little Pedro Ximenez is still being planted in Australia, although not enough to replace removals. For best results it needs a reliably dry period for ripening and harvest which Australian viticultural areas cannot provide.

Pedro Ximenez is a vigorous, upright and productive variety which, however, has a greater than average susceptibility to powdery mildew. The leaves are medium in size and readily recognisable, being deeply 5-lobed, fresh green in colour, thin and hairless below with only a few bristles along the veins. The bunches are large and well filled with medium, round to short oval berries, which have a tender skin and are very susceptible to rain damage. This can be very severe and followed by very extensive mould growth on vines with a dense canopy and a restricted number of heavy compact bunches. Vines at Merbein trained to a 90-cm wide T trellis and cane pruned have produced more exposed, smaller and looser bunches that are less susceptible to rain damage. Such vines provide conditions much less conducive to mould development.

In Spain the variety is used for both dry and sweet fortified wines such as Montilla-Moriles, Malaga and Jumila. Sweet wines may be produced either by adding alcohol before fermentation is finished or by adding mistelle (fortified grape juice) or grape juice itself to fully fermented wines. Wines are also made from grapes partially dried in the sun after harvesting. In Australia it provides excellent sherry material and in cooler areas it can also give good, fresh, neutral table wines suitable for blending with more highly flavoured varieties.

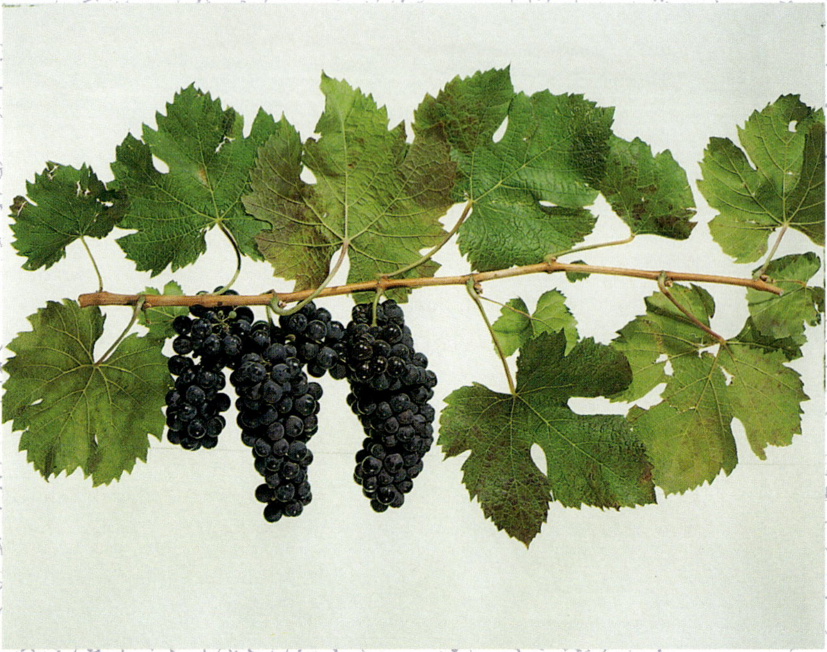

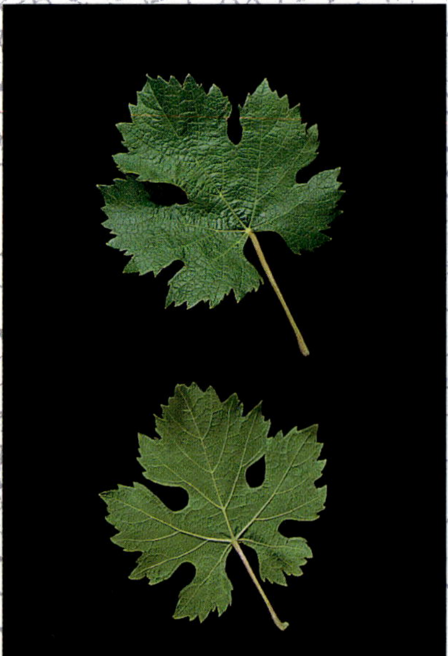

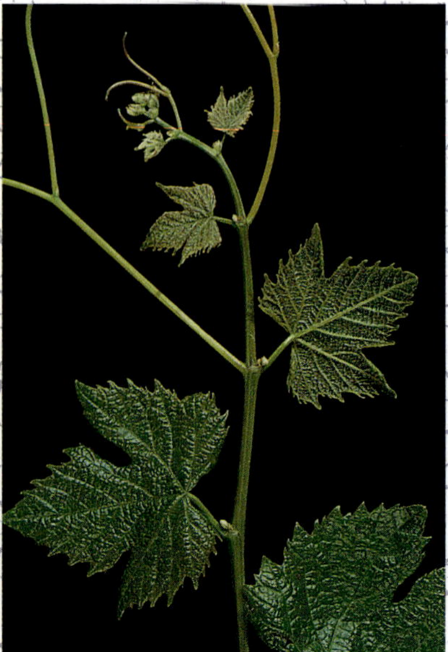

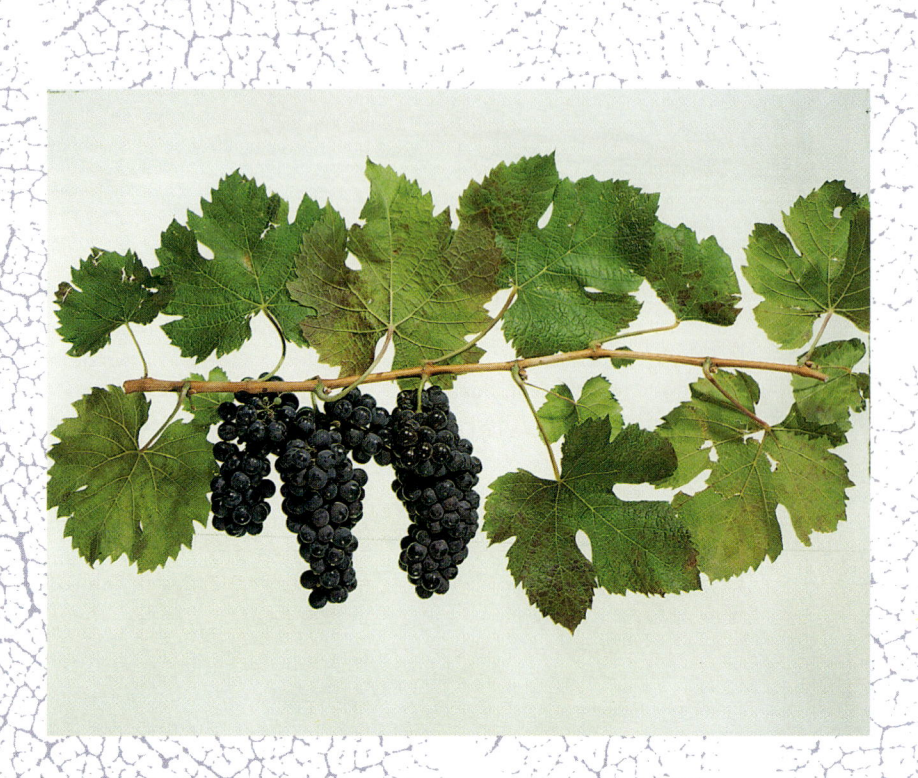

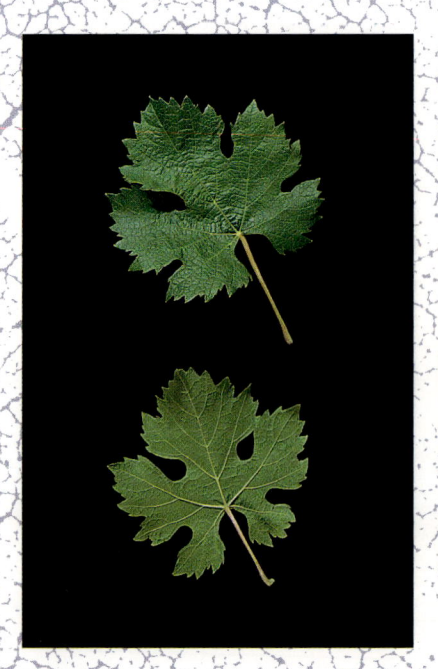

Peloursin

Peloursin is an old variety from the east of France, where it found favour with growers because of its hardiness and abundant yields. Its great vigour made it very suitable for pergolas around the home. However, it was never recommended or authorised when the varieties were classified and it has now almost disappeared; only 10 ha remained at the 1968 census. It has survived unrecognised in California and Australia, mixed with Durif in plantings called Durif in Australia and Petite Sirah in California.

Peloursin resembles Durif, although it is not difficult to tell established vines apart. Peloursin is more vigorous and, while it has a spreading habit of growth, it tends to be more upright than Durif. The leaves are large, more deeply 5-lobed, and hairless both above and below. The bunches are larger and more compact than those of Durif and susceptible to bunch rot. The berries tend to be larger and quite round unless pressed out of shape and are similar in their dark bluish-black colour with little bloom.

Peloursin was never considered suitable for using alone in wine, but was combined with a variety having more body and character. The variety originally used, Persan, does not seem to have been imported to California and Australia, unlike Durif which was not used until later, when it became established in France.

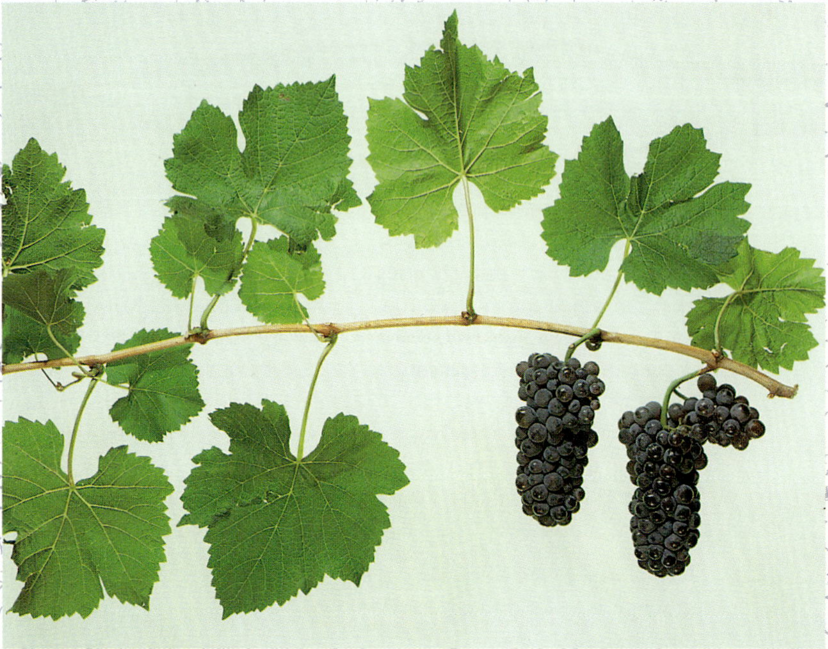

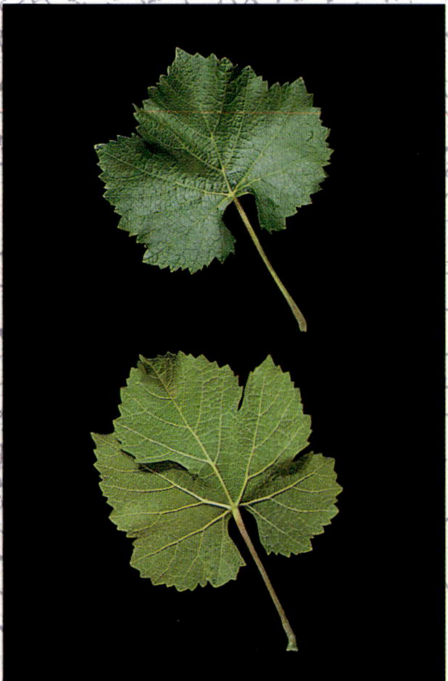

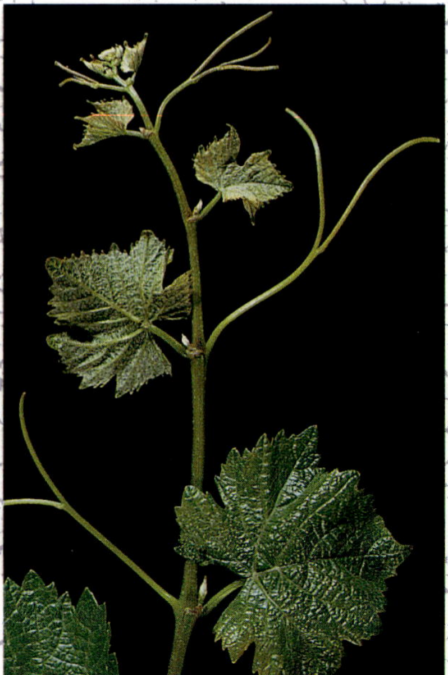

Pinot Gris

Like Pinot Noir, Pinot Gris is grown in many countries. There are about 900 ha in France, mostly in Alsace, 2500 ha in Germany where it is known as Rulander, and probably something of the order of 2200 ha in northern Italy. It is also grown throughout central and south eastern Europe, one of its more interesting names being Szurkebarat (Grey Friar) in Hungary. It has not attracted much interest in California or Australia.

Pinot Gris differs from Pinot Noir only in having much less pigment in the skin of the berries. It is used to make a deep golden wine which may at first sight suggest an oxidised white wine. However, in the form of such wines as the Tokay of Alsace, it is highly regarded in Europe.

Pinot Blanc differs from Pinot Noir and Pinot Gris only in having no pigment in the skin of the berries. It makes a distinctive wine but with less varietal character than Riesling or Chardonnay.

Pinot Blanc is probably the least grown of any of the Pinot family. The situation has been confused because Chardonnay has often been called Pinot Chardonnay or even Pinot Blanc Chardonnay and so has come to be called Pinot Blanc in some areas. In South America Chenin Blanc has been wrongly identified as Pinot Blanc, while in California the variety Melon has been wrongly called Pinot Blanc. The true Pinot Blanc has only recently been imported into Australia and earlier plantings so called would be either Chardonnay or derived from the Californian Melon. There are about 1500 ha of true Pinot Blanc in France, mainly in Alsace, possibly about the same in northern Italy and a little in Germany and central Europe.

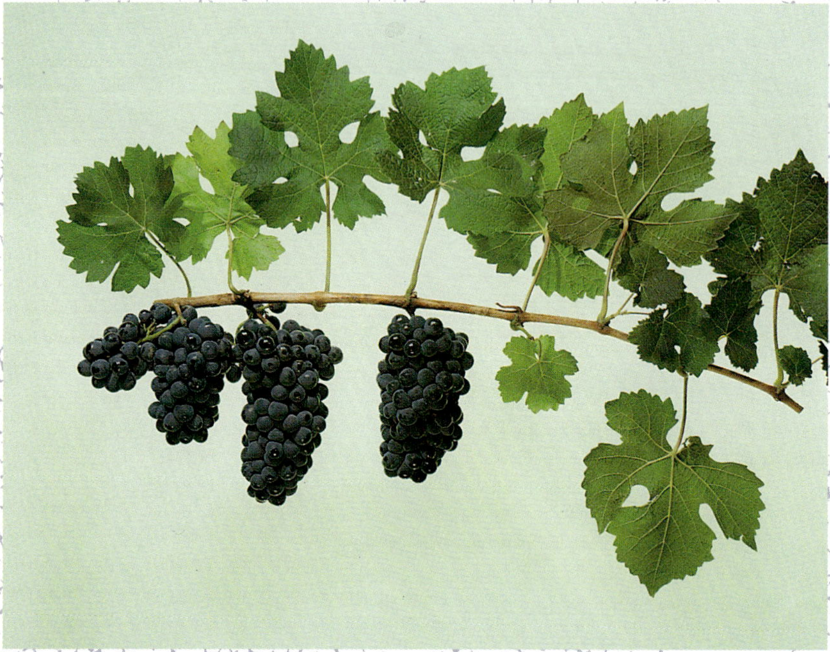

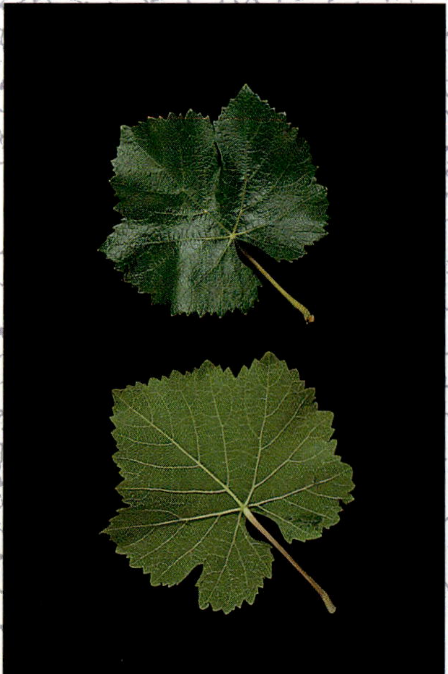

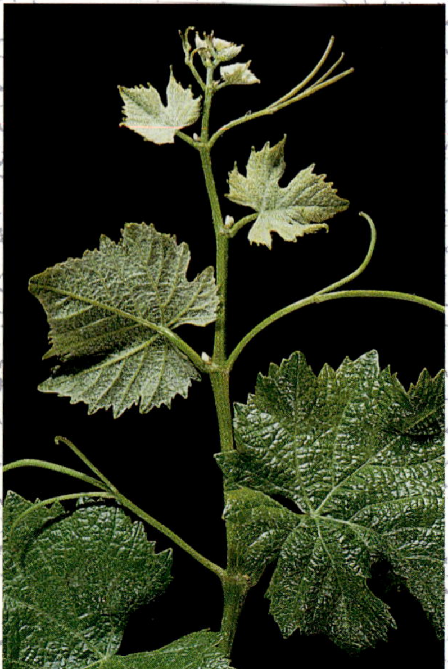

Pinot Noir

Pinot Noir is the variety used in the superior red wines of Burgundy and one of the principal Champagne varieties. Plantings in France increased from 8500 ha to 17 300 ha between 1958 and 1979 and reached 22 000 ha by 1988. It is one of the few black varieties ripening early enough to succeed in the coolest viticultural areas such as those of Germany and Switzerland. It is grown in practically every country in which wine grapes are grown, usually not extensively and only in the cooler areas such as the north of Italy or the coastal valleys of California. By 1992 more than 3700 ha had been planted in California and more than 1300 ha in Australia.

Pinot Noir is a very old and variable variety and a progression can be seen from vines similar to the wild grapes that grew in Europe before cultivation to high-yielding selections sometimes thought to show less varietal character in the wine. There are about thirty different recognised clones in Australia with observable differences in growth habit, bunch shape and so on. It is quite possible that some clones will be better adapted to particular areas and this will need to be taken into account when evaluating the variety.

Pinot Noir is not vigorous and takes some years to develop vines of a reasonable size. Some clones show more upright growth than others. The leaves are small to medium, usually almost entire, but sometimes deeply 5-lobed on water shoots or unfruitful vines. They are rough, tend to fold inwards about the midrib and have a few tufts of hair on the lower surface. The bunches are small, cylindrical and winged, and because they are very compact the small, short oval berries are often pressed out of shape. Several clones with less compact bunches have been selected in Switzerland for use in areas where bunch rot is a problem. Even for selected clones yields are only moderate.

The colour of wines from Pinot Noir is never intense and fruit from hot areas may make uninteresting wines lacking in colour and flavour. However, in cool areas the wines have a distinctive varietal flavour which is highly esteemed.

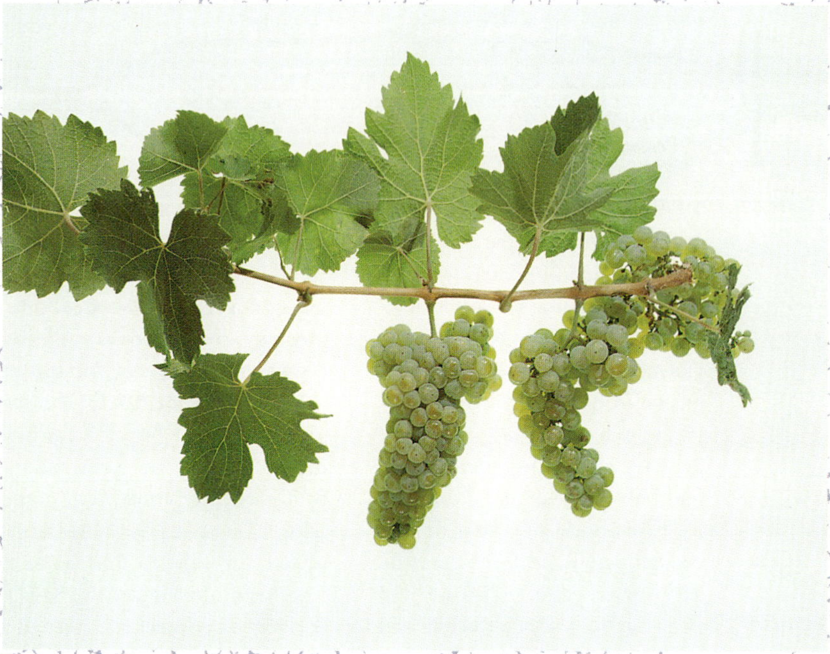
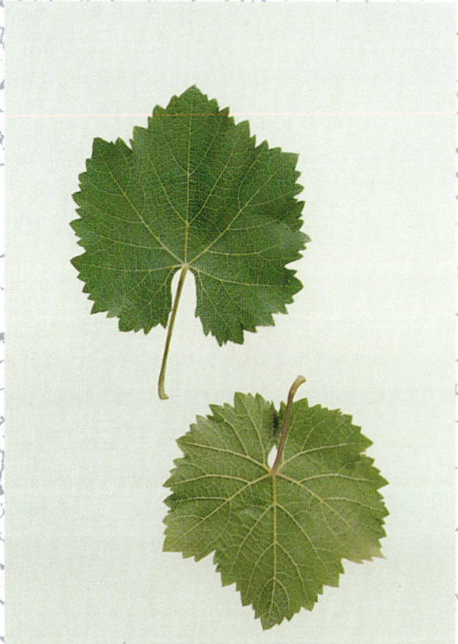
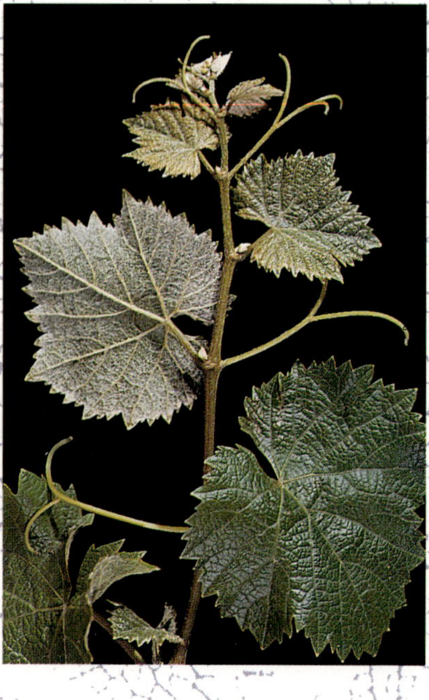

Riesling

Riesling is the noble wine grape variety of Germany. In 1994 there were 23 000 ha in the German Federal Republic, second only to the 25 000 ha of Müller-Thurgau. Although often regarded as the standard by which white wine varieties are judged it is not planted to the same extent in any other country. The area of Riesling in France had increased from 1074 ha in 1968 to 2918 ha in 1988. There are modest areas in northern Italy, and in the eastern European countries. There are only about 50ha in Argentina and very little in Chile and South Africa. Most of the so-called Riesling in South Africa appears to be Crouchen. California with 1700 ha and Australia with 3600 ha may well have the largest areas outside Germany and France.

In Australia it is often called Rhine Riesling to avoid possible confusion with Hunter River Riesling (Semillon) and Clare Riesling (Crouchen). This tendency to use the name Riesling for other varieties is not confined to Australia and similar distinctions are needed in other countries. Thus it is called Riesling Renano in Italy, Rheinriesling in Austria, Rajnai Rizling in Hungary and Rajnski Rizling in Yugoslavia mainly to distinguish it from the variety known as Riesling Italico, Welschriesling, Olaszrizling or Rizling Vlassky respectively which provides much of the Riesling wine from these countries. In California it is called White Riesling, to separate it from Grey Riesling, a grey-fruited form of Bastardo, and there are also such names as Frankenriesling (Sylvaner), Breisgauer Riesling (Ortlieber), Budai Rizling (Kleinweiss) and Banati Riesling (Creaca).

In the hot areas of Australia Riesling does not show the vigour found in cooler climates and is a variety of only moderate vigour. It has rough, dark green leaves, entire to 3-lobed with a few tufted hairs on the lower surface. The leaf stalks usually show some red colouration which may continue into the base of the veins. The bunches are small and compact with short and often woody stalks. The berries are small and round, gold where exposed to the sun and irregularly marked with brown spots, and juicy with a tough skin.

Riesling has a definite but not overpowering varietal character which is maintained over a wide range of climates. It can show up well in dry or slightly sweet wines made from sound grapes, in luscious sweet wines made from grapes affected by noble rot or from the concentrated juice separated from the ice of grapes partially frozen by severe frost.

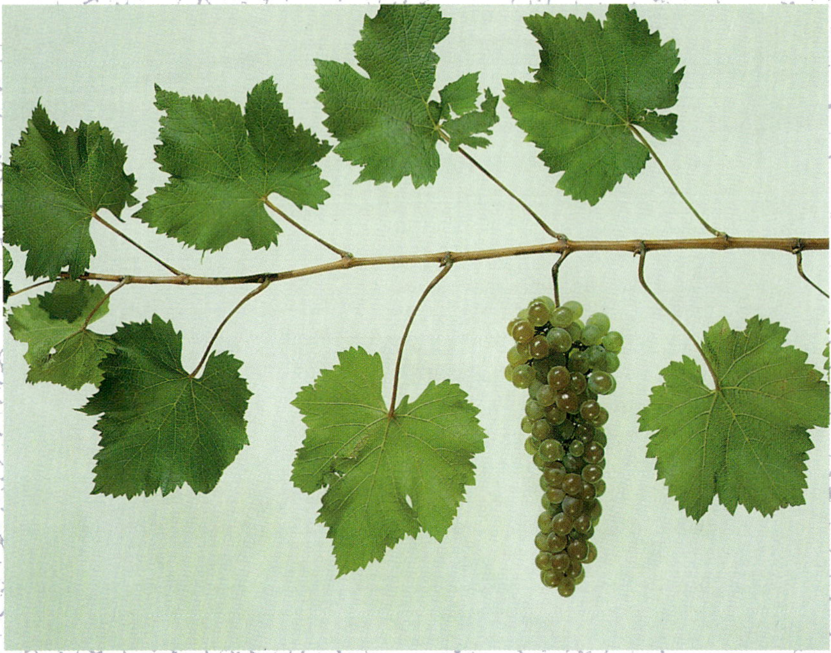
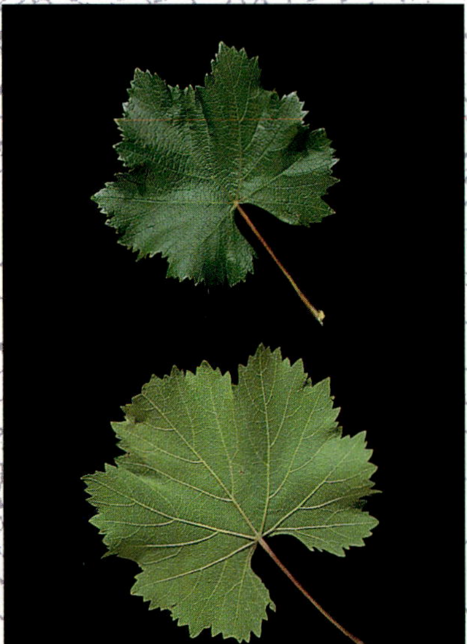
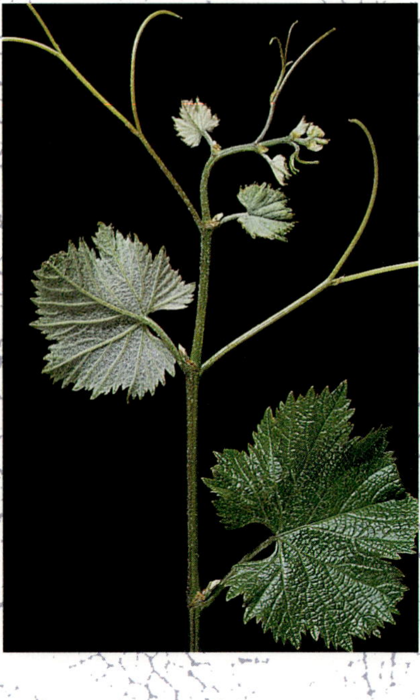

Rkaziteli

Rkaziteli is a local variety from the Tbilisi region of Georgia where it is grown around the town of Telavi in the Alazani river valley. It is also grown in Moldavia on the Black Sea coast where it is known as Gratiesti and in Bulgaria where the wine is exported to Germany under the name of Sonnenküste. It was imported into Australia in 1971 and a small amount has been planted in the warm irrigated region of the Murray River Valley.

It is a vigorous variety with a strong erect growth habit. It is not fruitful in the basal buds and should be cane pruned. The growing tip is felty white with strong carmine colouration on the margins of the young leaves. Shoots are long and straight, green with reddish nodes and turning yellowish brown when mature. Leaves are medium in size, a little longer than wide and 5-lobed. The petiolar sinus is lyre shaped and is sometimes closed. The leaves are dark green with some blistering on the upper surface. They are glabrous on the upper side with some tufted hairs underneath. The teeth are irregular and obtuse. The bunch is medium in size, conical, elongated and loose with a long peduncle. The berries are white, medium in size, slightly ovoid with a thin skin which often shows a reddish golden colouration when exposed to the sun.

In Australia the yields are medium to low. The wines are well balanced but generally lack character.

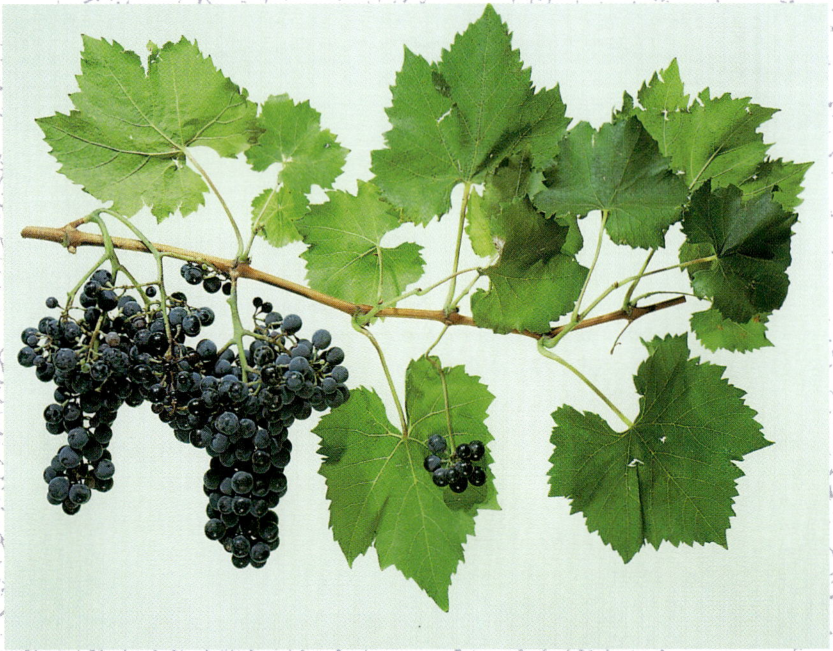
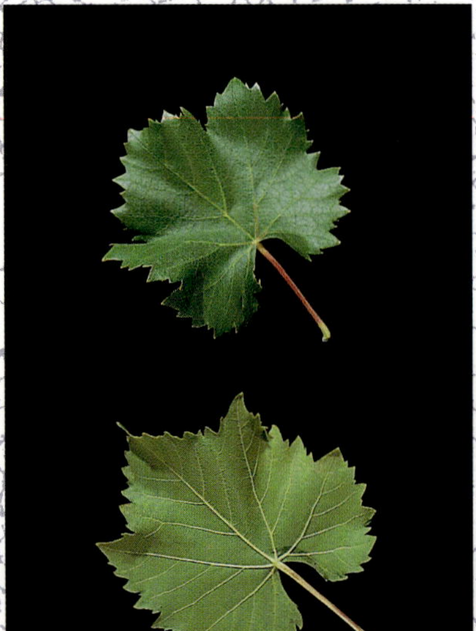
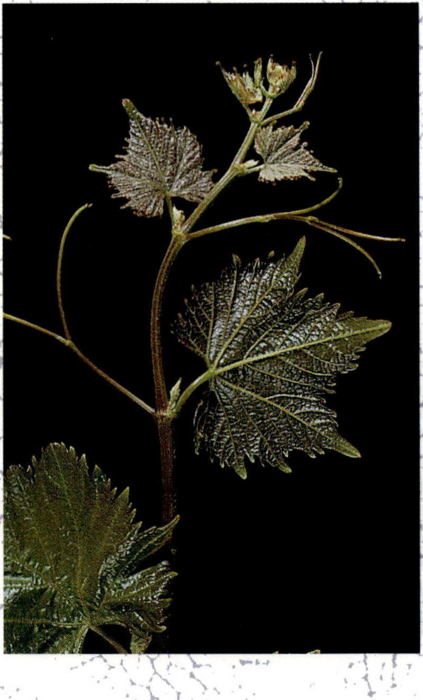

Rubired

Rubired is another new variety from H.P. Olmo, released in 1958. It is a cross of Tinta Cao, a port variety, and Alicante Ganzin, which has an Aramon x Rupestris rootstock as one parent. Thus Rubired has one-eighth of the wild American species *Vitis rupestris* in its make-up. This has some advantages, such as a degree of resistance to fungus diseases, but also leads to some problems, such as dense foliage and numerous light bunches. Plantings in California reached 5300 ha by 1974 but had declined to 2773 ha by 1992. A few fairly substantial plantings were made in Australia around 1970 but they have been largely grafted over or removed.

Rubired is a vigorous variety with a spreading habit of growth. It has medium to large leaves, entire to slightly 3-lobed, tending to roll back a little at the edges, with the leaf stalk often unusually long in relation to the size of the leaf blade. Both leaf surfaces are hairless, but the upper surface is a much darker green than the lower. Bunches may have a fairly large framework but are loose to very loose and so medium or less in weight. Berries are small and nearly round with red juice.

Wines from Rubired are opaque with a very intense deep red colour. They have an unusual aroma which is often found in wines made from varieties with red juice but are neutral enough to be blended in small quantities to improve the colour of wines deficient in this respect.

A related variety, Royalty, a cross of Bastardo and Alicante Ganzin, was released by Dr Olmo at the same time as Rubired. It is reported to be more demanding as to soil type, less vigorous and lower yielding. Originally planted in California to about the same extent as Rubired it did not show the same expansion, reaching only 1200 ha by 1974 and declining to 340 ha by 1992. Both varieties were originally released for making fortified wines of the port type and Royalty appears to be preferable for this purpose.

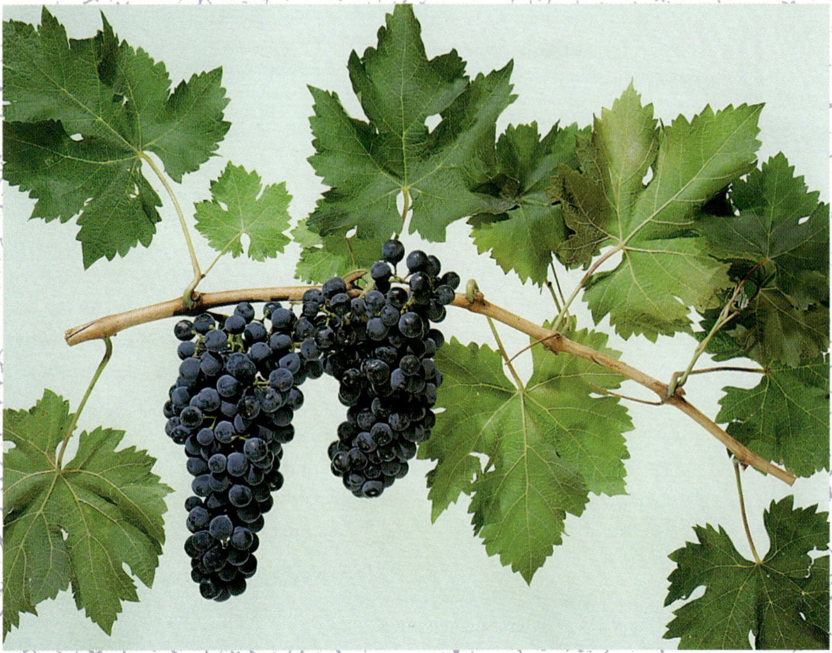
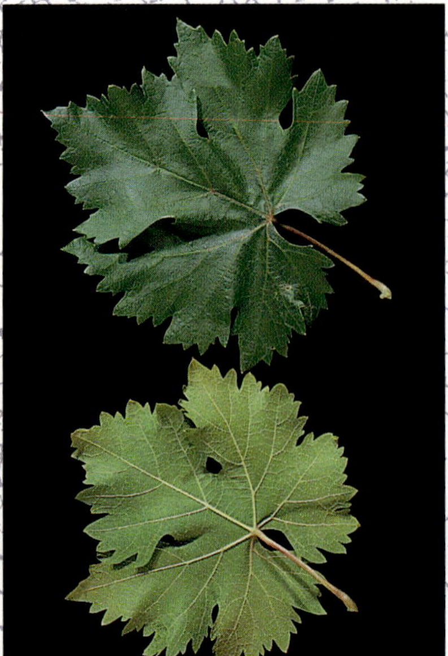
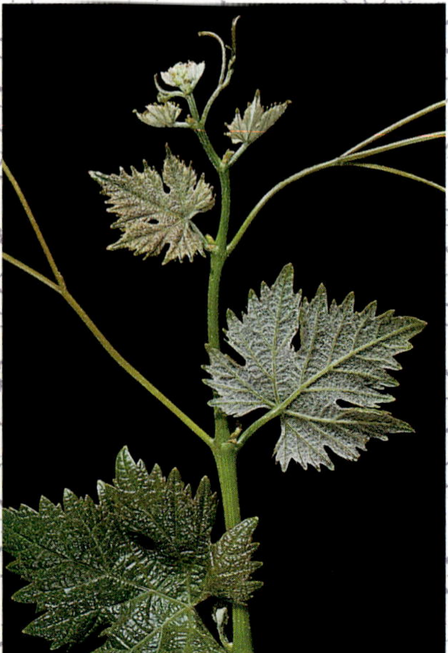

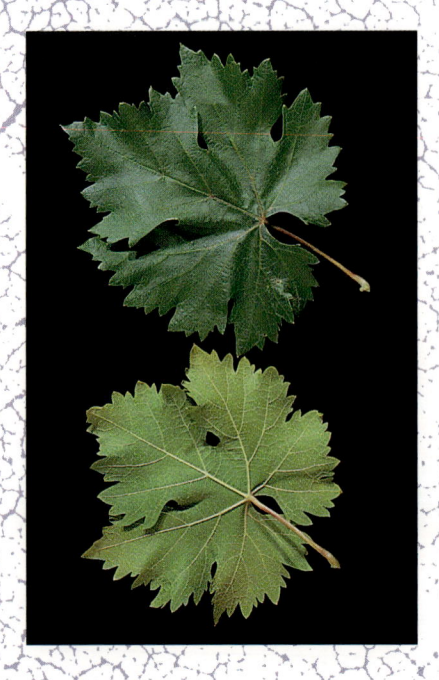

Ruby Cabernet

Ruby Cabernet, a cross between Carignan and Cabernet Sauvignon, is a variety bred by H.P. Olmo at the University of California, Davis, and released in 1948. Its major commercial exploitation has been in California, where 8064 ha had been planted by 1977, but the area had declined to 2643 ha by 1992. A total of 470 ha had been planted in Australia by 1992.

Ruby Cabernet shows some characters from each of its parents. It is vigorous and tends towards the upright growth habit of Carignan. It has large, 5-lobed leaves more like those of Carignan than those of Cabernet Sauvignon. The bunches are medium, loose and generally winged, and the short oval berries approach those of Carignan in size. The bunches have woody stems which make them difficult to harvest by hand and major planting in California took place after mechanical harvesting became available. Ruby Cabernet is only a little less sensitive to fungus diseases than Carignan.

The variety was designed for producing high quality red wines in the hot dry areas of California where Cabernet Sauvignon is not recommended. In this it appears to be successful—it has produced some excellent wines at Merbein although the wines may mature more quickly than those of Cabernet Sauvignon. At Merbein, Ruby Cabernet ripens at the same time as Cabernet Sauvignon and it has not shown any difference in sugar or sugar– acid ratio to suggest that it should be preferred on this account.

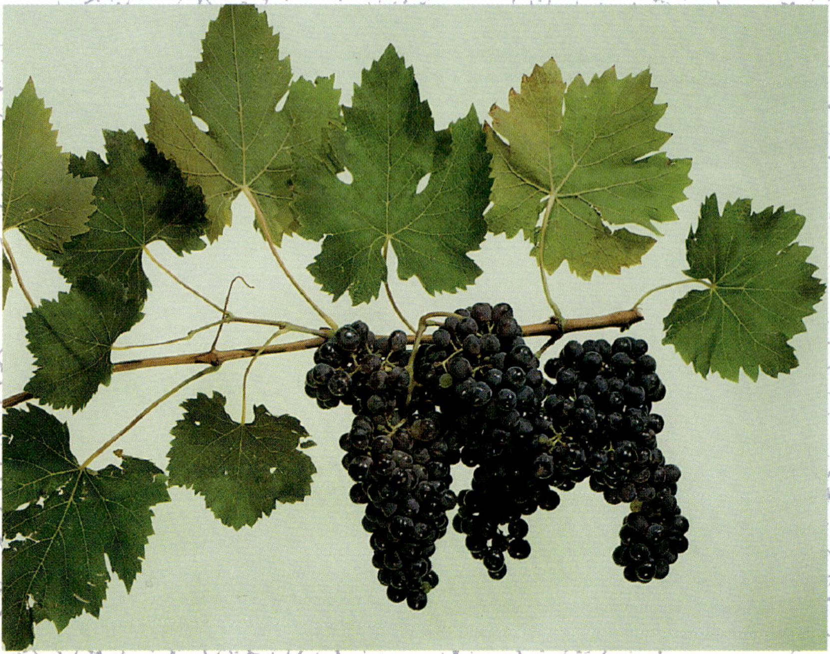
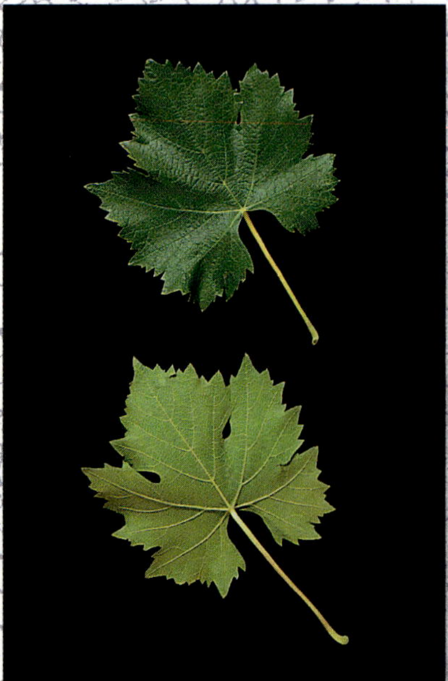
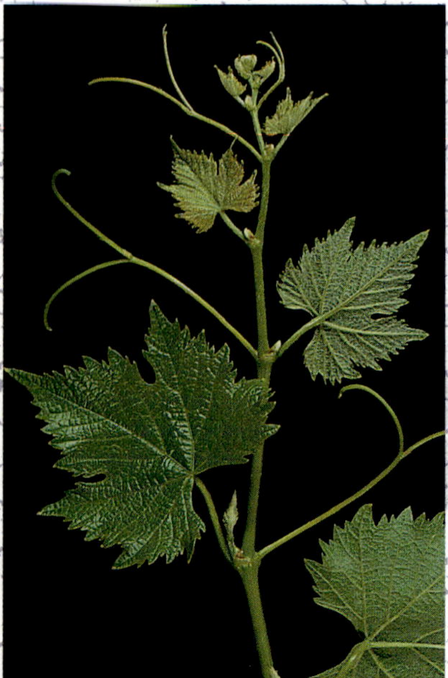

Sangiovese

In terms of area planted and grapes produced, Sangiovese is the leading wine grape of Italy. Because of mixed plantings (trees and vines) the area is not well defined but something like half a million tonnes of grapes would be produced each year. The variety is thought to be originally from Tuscany but it is also recommended for many other parts of Italy. Apart from 5000 ha in Argentina, Sangiovese does not appear to have been planted much in other countries, and in Australia was found only as odd vines at Mudgee under the name of Canaiolo, which is another Tuscan variety.

Sangiovese is a fairly vigorous variety with a spreading habit of growth. Leaves are medium, nearly entire to 3-lobed, almost smooth and flat, with a few tufted hairs on the lower surface. The petiolar sinus is an open U. Bunches are small to medium, winged and well filled to compact. Berries are medium and short oval, with a heavy bloom.

Sangiovese is used in many wines of controlled appellation in Italy, sometimes alone, but quite often mixed with three or four other varieties. Perhaps the best known of these wines would be Chianti, for which 50–80% of the grapes used are Sangiovese, along with Canaiolo, Trebbiano, Malvasia del Chianti and a little Colorino.

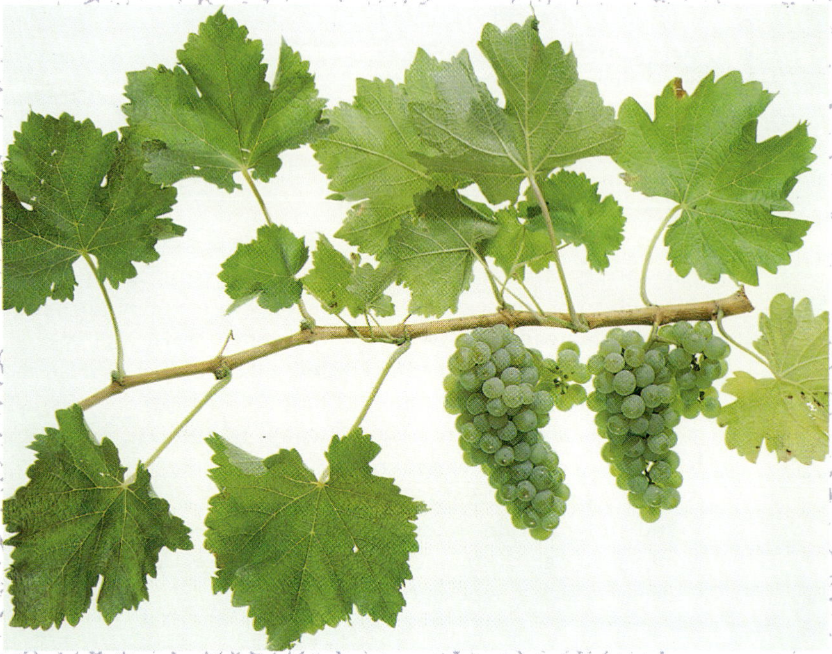

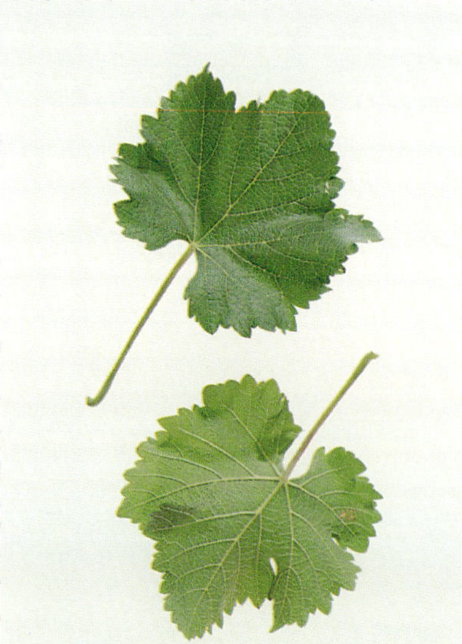

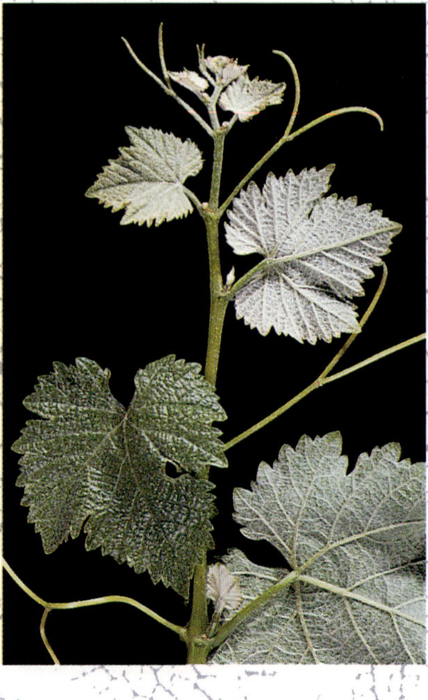

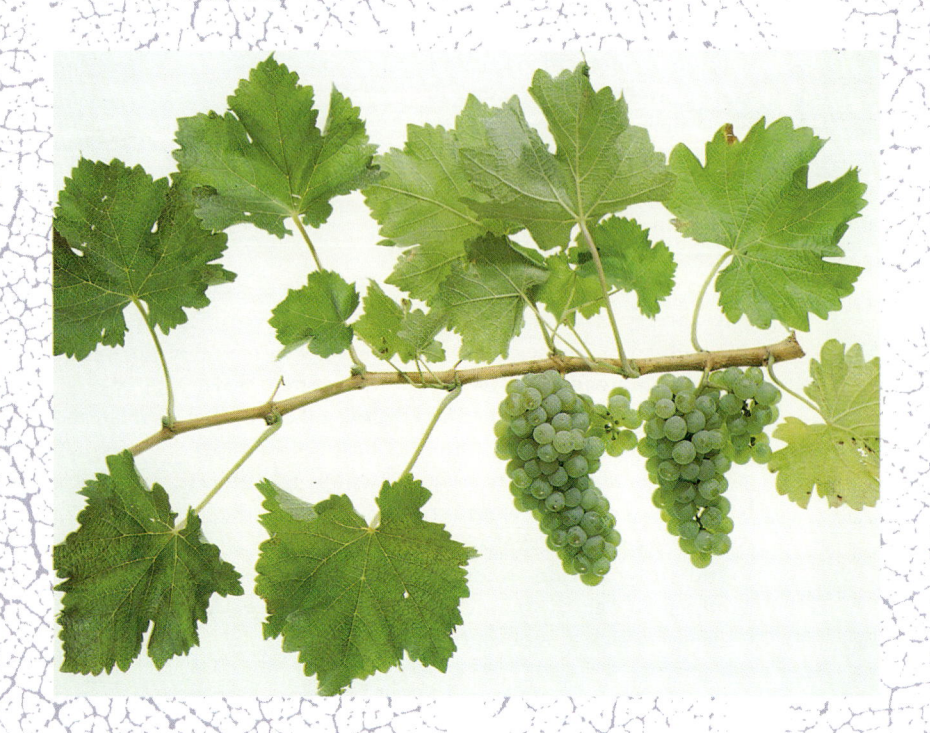

Sauvignon Blanc

With a total of about 12 000 ha in 1988, Sauvignon Blanc ranked fifth among the white wine grape varieties of France. The area had increased considerably since 1958, when Sauvignon Blanc ranked eighteenth. The main plantings are near Bordeaux, where it is used as a minor but important partner of Semillon. There are smaller plantings in the Loire Valley, where it is used on its own. There would be a few thousand hectares of Sauvignon Blanc in northern Italy and it has also spread to eastern European countries. California has about 5400 ha, but most of the 3400 ha of so called Sauvignon in Chile are a closely related variety known as Sauvignonasse in France and Tocai Friulano in Italy. In Australia there are about 1000 ha of Sauvignon Blanc, rather more than half in South Australia and the rest divided between Victoria, New South Wales and Western Australia.

Sauvignon Blanc is a vigorous and rather upright variety. It has rough, undulating 3- to 5-lobed leaves which differ from those of Semillon in having more rounded teeth, a little more hair on the lower surface, a somewhat brighter green colour and a more frilly appearance. The variety Sauvignonasse mentioned above, which is present in Australia as odd vines in plantings of other varieties, has leaves of an even lighter green and a very frilly appearance. Sauvignon Blanc has small, cylindrical, compact bunches which are sometimes winged, and small, short oval, greenish-yellow berries

In a cooler climate, such as that of the Loire Valley, Sauvignon Blanc gives wines with a very strong varietal character. This is also found in wines from the coolest Australian areas and its unfamiliarity may hinder immediate acceptance of the variety. The varietal character is less strongly developed in warmer areas, where Sauvignon Blanc gives pleasant, fresh, acid wines. It is considered a very desirable component in the white wines of Bordeaux, and similar mixtures of Semillon and Sauvignon Blanc probably deserve more extensive trialing in Australia.

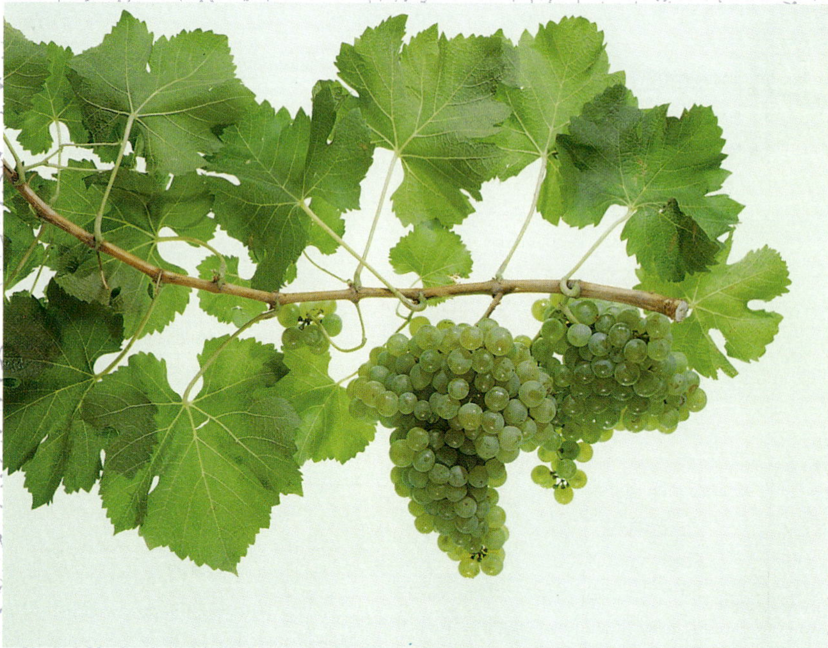
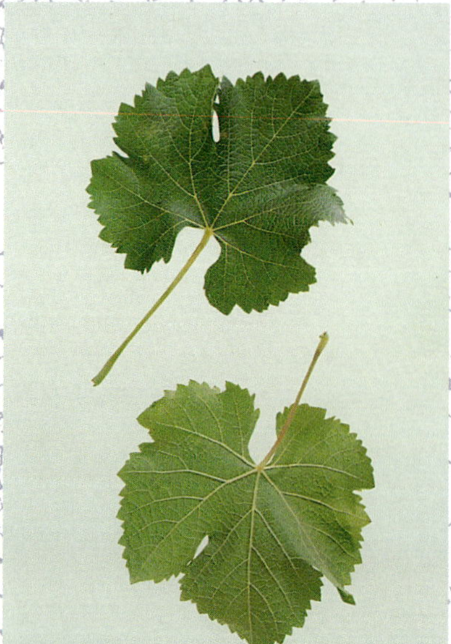
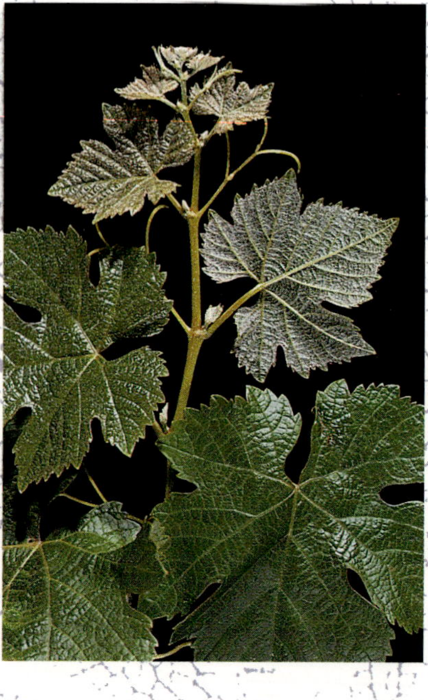

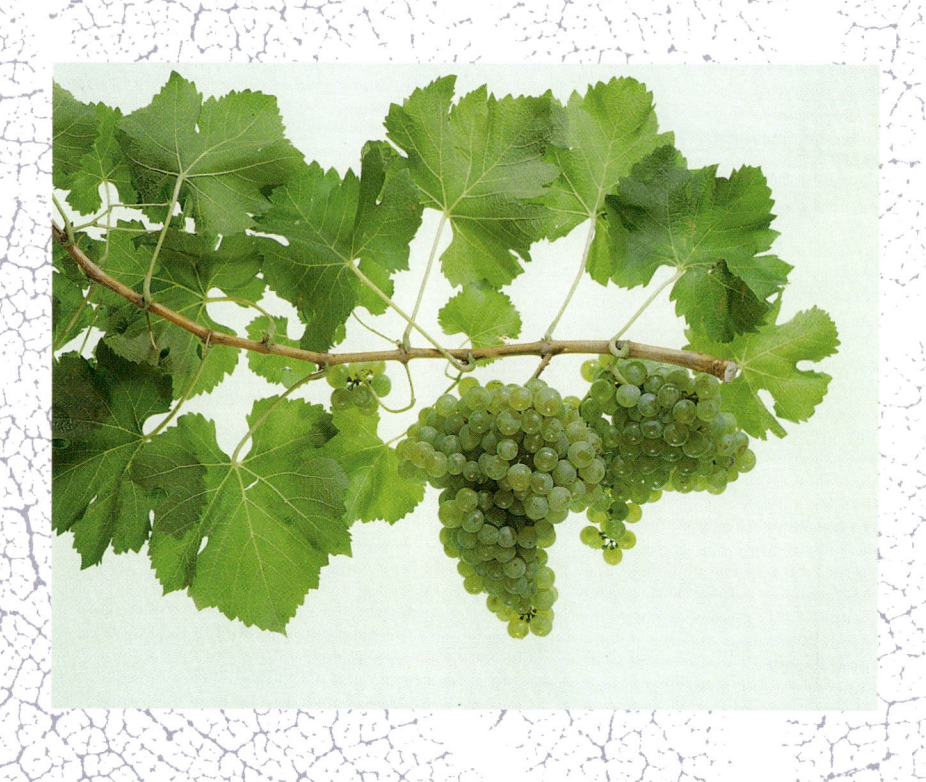

Semillon

Semillon is the major white variety of the Bordeaux area of France, and ranks third in France as a whole with 17 500 ha. A little is planted in Italy and Yugoslavia and perhaps more in the Soviet Union. California has about 840 ha but the main plantings outside France are in the Southern Hemisphere. Semillon is the leading white variety in Chile with about 35 000 ha. There are about 2500 ha in Argentina and 1000 ha in South Africa where it is known as Greengrape. In Australia there are nearly 3000 ha, about three-quarters in New South Wales and most of the rest in South Australia. It proved to be well suited to the Hunter Valley and became known as Hunter River Riesling.

There has been much confusion between Semillon and several other varieties in Australia. The variety known as Hunter River Riesling in Western Australia was found to be Chenin Blanc. In South Australia several hundred hectares of Crouchen were planted erroneously as Semillon and another small planting thought to be Semillon was Riesling while a true Semillon was present under the name of Madeira and as one small planting called Sercial. 'Barnawartha Pinot' of north east Victoria is also Semillon.

Semillon is a fairly vigorous variety with clones varying from spreading to rather more upright growth. It has a characteristic rough, undulating, 3- to 5-lobed leaf with a few tufted hairs on the lower surface. Bunches are medium, conical, and well filled to compact. Berries are rather small, round to short oval, and juicy, with only fair resistance to rain damage. Only the white-fruited form of Semillon is known in France, but in both Australia and South Africa red-fruited forms have been found.

In France, Semillon is often affected by noble rot which concentrates the berry constituents and modifies the flavour. Such grapes are used in the luscious, sweet wines of Sauternes and Barsac, while unaffected grapes are used for dry wines. In both cases the Semillon is fermented together with a smaller quantity of Sauvignon Blanc and often a little Muscadelle to add further desirable flavour components. In Australia, Semillon is used for dry white wines which, particularly when yields are not too great, have a distinctive varietal character which lends itself to aging.

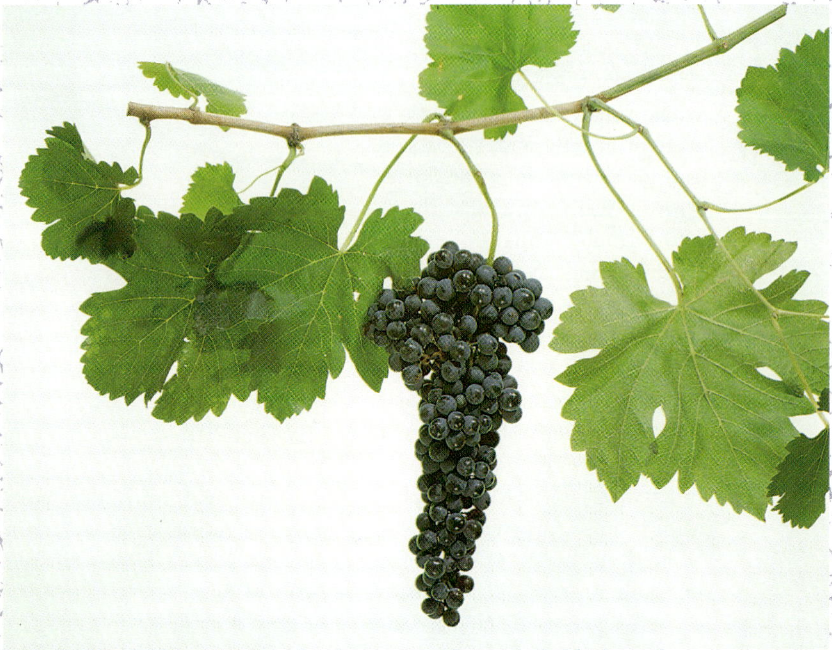

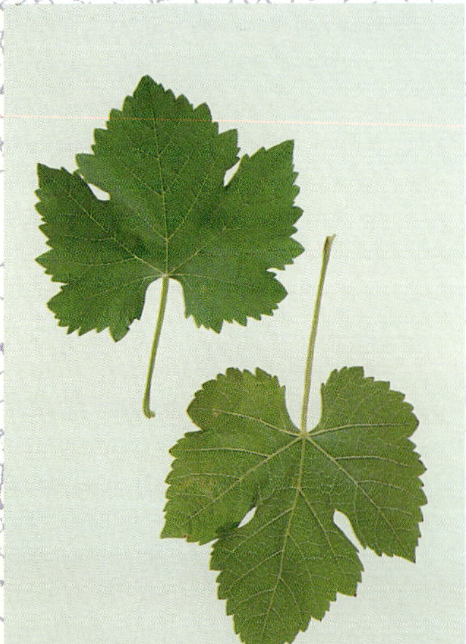

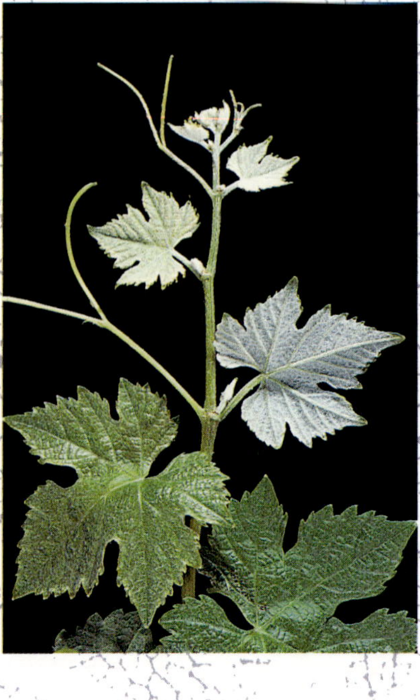

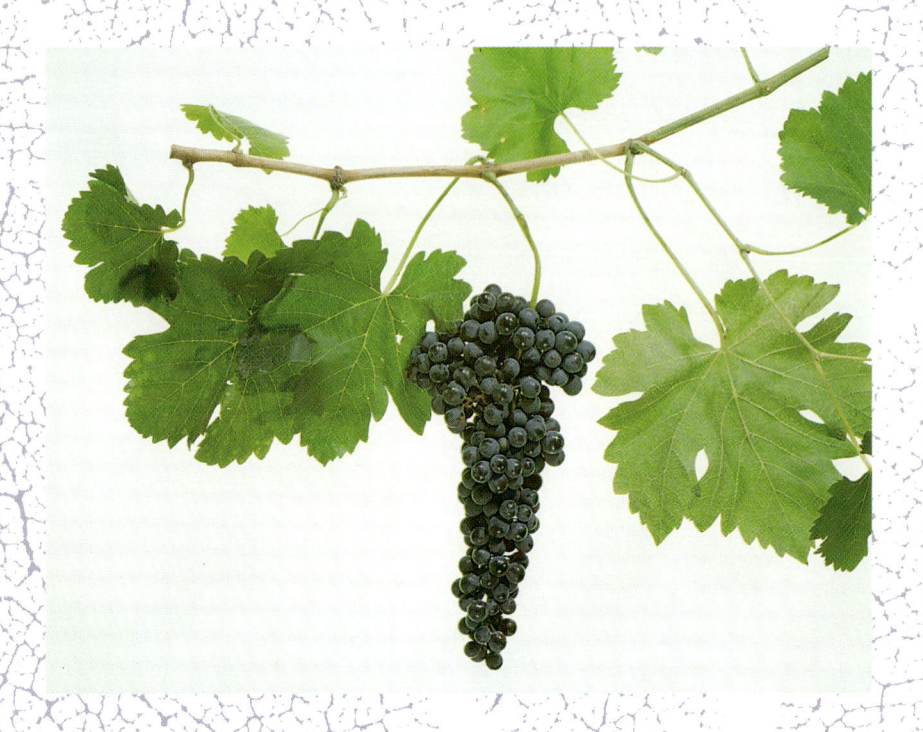

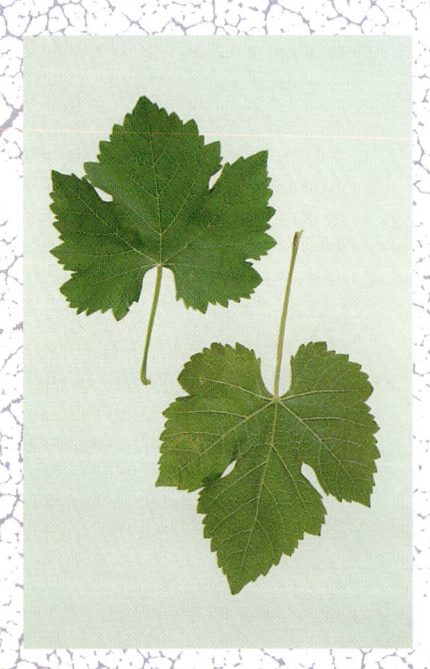

Shiraz

This variety comes from the Hermitage area of the Rhone Valley in France where it is known as Syrah. One tradition suggests that it was brought to Hermitage from Shiraz in Iran by the hermits, another that it was brought from Syracuse by the Roman legions, but it seems quite likely that it originated in the Rhone Valley. It is sometimes called Hermitage in Australia, but should not be confused with the Hermitage of South Africa which is really Cinsaut. The Petite Sirah of California is also a different variety, Durif, which comes from the same part of France where it has occasionally been incorrectly called Petite Syrah. There has been a dramatic increase in the area planted to Syrah in France, from 2700 ha in 1968 to 27 000 ha in 1988, mainly to add character to wines based on Grenache or Carignan. There were just over 6000 ha of Shiraz in Australia in 1992. A little is grown in Tuscany in Italy and over 1000 ha in Argentina. There are about 500 ha of true Shiraz in South Africa.

Shiraz is a vigorous variety with a spreading habit of growth. It has medium, 5-lobed leaves, somewhat rough and undulating, with tufted hairs on the lower surface. The bunches are characteristically long and cylindrical with long stalks, rather loose, with small to medium oval berries which tend to wilt as soon as they are ripe, becoming more difficult to harvest mechanically.

In Australia, Shiraz has proved to be a very versatile variety. It is grown in all viticultural areas and used for all types of red wines. It is sometimes used alone but is often blended with other red varieties.

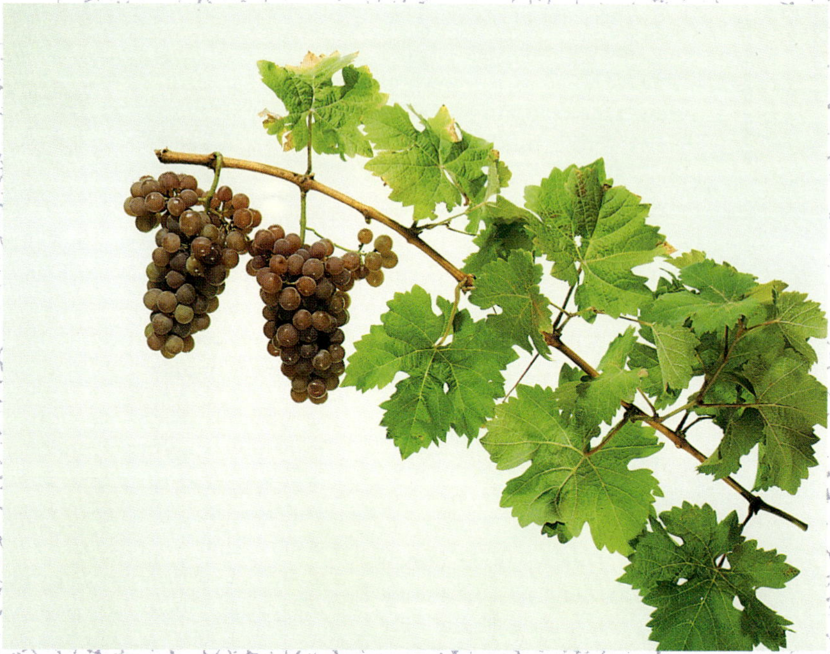

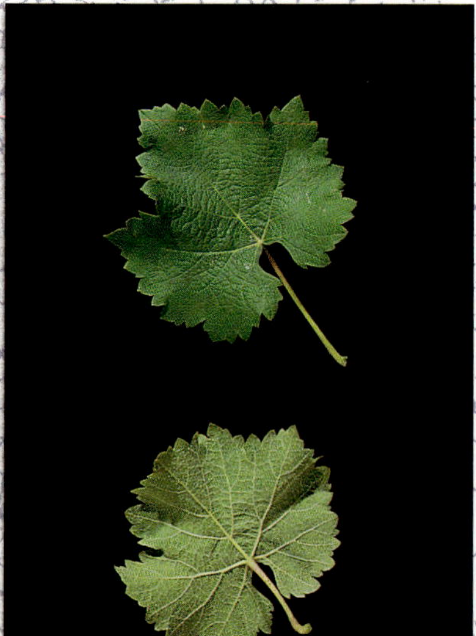

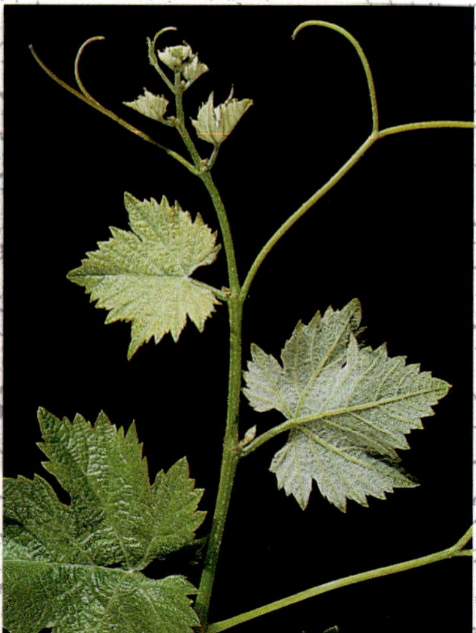

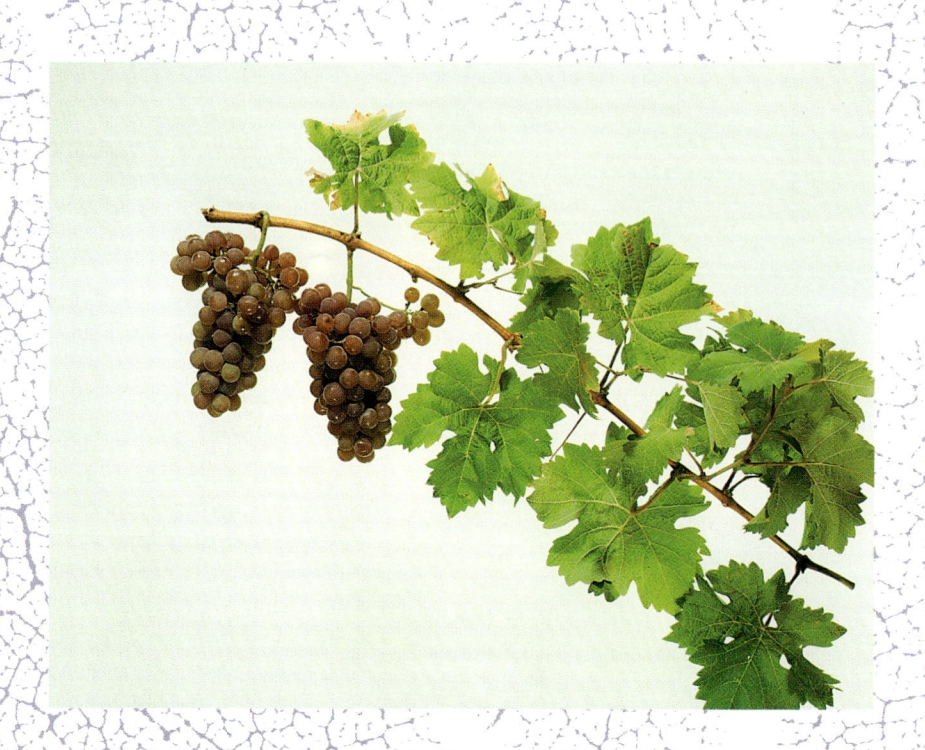

Siegerrebe

Siegerrebe is a variety bred at the Grape Breeding Station at Alzey, Rheinland-Pfalz, Germany and released in 1958. It is a cross of Madeleine Angevine x Gewürztraminer. An attractive pink grape, it is amongst the earliest ripening of all the white winegrape varieties. It has not been widely planted in Germany with only 269 ha recorded in 1981.

Siegerrebe has an open growing tip which is covered with dense hairs. The shoot tip is white to yellowish green with a reddish tinge. The leaf is small to medium in size, 3- to moderately 5-lobed with a wide middle lobe. The surface is bullate with blunt teeth. The petiolar sinus is closed and overlapping. The mature canes are yellowish brown in colour. The bunch is medium in size and loose. The berries are round, medium in size and pink when mature with strong, spicy varietal character. The variety has medium to strong vigour and both budburst and maturity are early.

Siegerrebe ripens in late January in the warm irrigated regions of Australia but has low titratable acidity levels. Yields are low, around 11 tonnes per hectare. The variety has performed well in cooler climates such as Tasmania where it ripens in mid February.

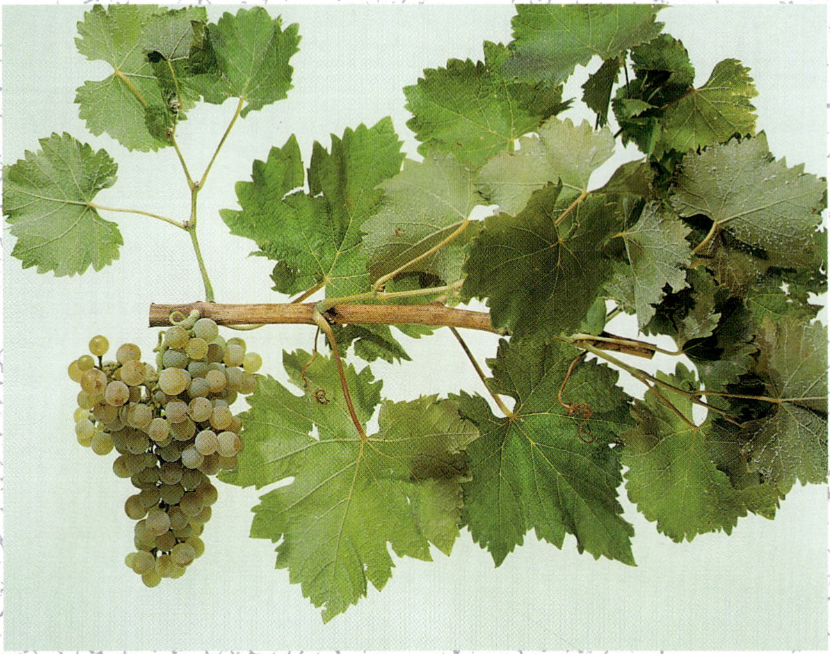

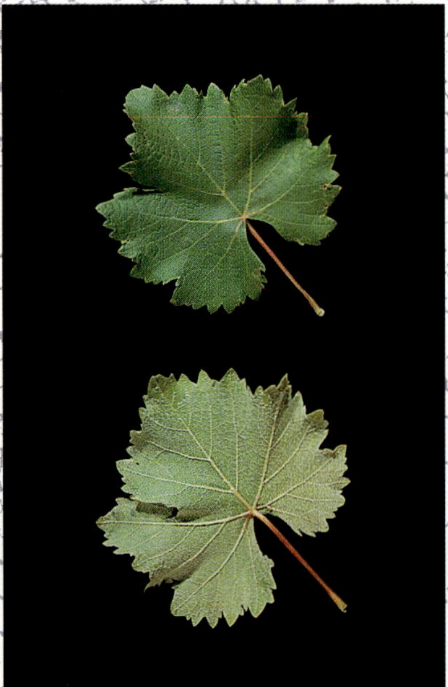

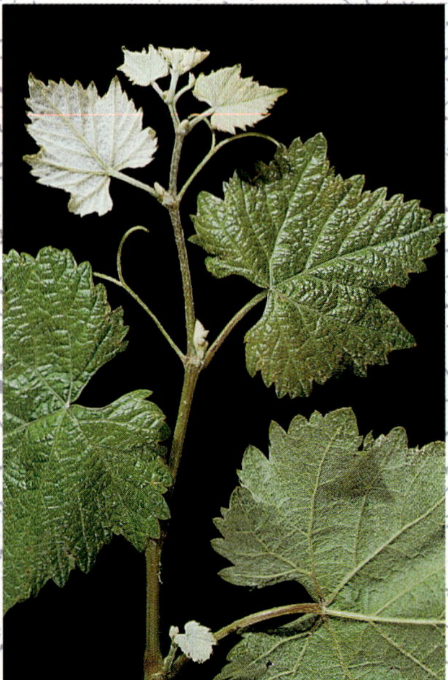

Solvorino

Solvorino is another variety which is known only from Australia at this stage. It has clearly been in Australia for well over 100 years and may have been one of the varieties introduced by Busby in 1832. Two varieties bred as table grapes have been named Solferino, a white in France from about 1850 and a red in Italy from about 1920. Both are thus too late to be connected with the Australian variety. Solvorino is grown commercially at Roma in Queensland and also occurs as previously unidentified vines in mixed plantings on two vineyards at Great Western in Victoria. One of these was planted in 1868 and can be traced to earlier plantings in Australia.

Solvorino is a vigorous variety with an upright habit of growth. Leaves are medium, entire to slightly 3-lobed, flat or tending to fold inwards, slightly rough with tufted hairs on the lower surface. The petiolar sinus is an open V and both the petioles and bunch stalks tend to be red where exposed to the sun. A characteristic feature is the way in which yellow areas develop between the main veins of leaves far out on the shoots and finally become rows of small holes. Bunches are medium, conical and compact and the berries medium and round with a heavy bloom.

Although Solvorino has no pronounced varietal character, it appears to be suitable for good standard wines such as those produced from Chenin Blanc and Colombard.

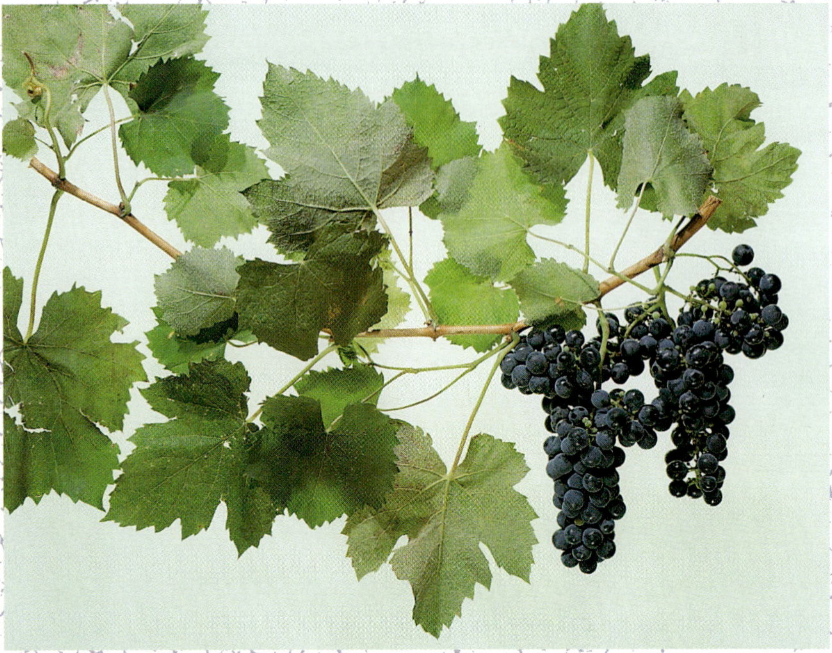

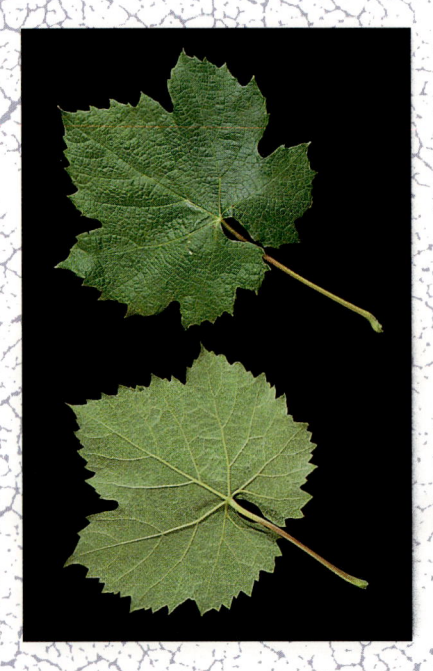

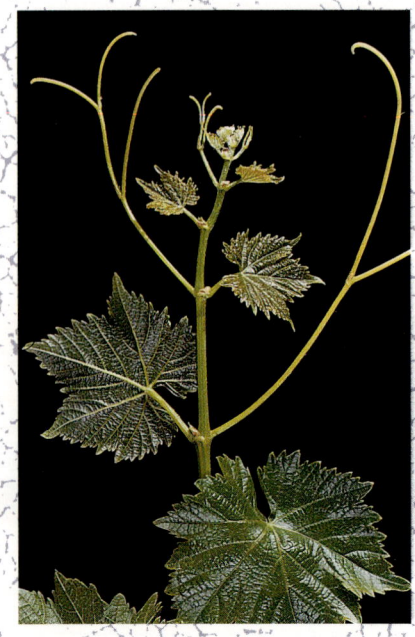

Souzao

Souzao is a Portuguese variety, this name being used for it in the Douro Valley where it is ranked in the second highest quality group for producing port. It is also grown in the Minho region where it is known as Vinhao, and to a slight extent in the neighbouring areas of Spain under the name of Souson.

Souzao is a vigorous variety with a trailing habit of growth. The leaves are medium in size, thin, slightly 3- to 5-lobed, tending to roll back a little at the edges, a little puckered near the point of attachment of the petiole and with a few hairs on the lower surface. The petiolar sinus is of a V or somewhat angular U shape. Bunches are small to medium, cylindrical and compact with medium, round berries. The variety shows a less than average susceptibility to downy and powdery mildews.

The main contribution of Souzao to the wines in which it is used is its excellent colour. It is equal or superior to some varieties with coloured juice such as Alicante Bouschet or Grand Noir de la Calmette in this respect. It also retains a good acidity. For both ports and dry red wines (vinhos verdes) in Portugal it is used in conjunction with other varieties which provide the other desired characters in the wine.

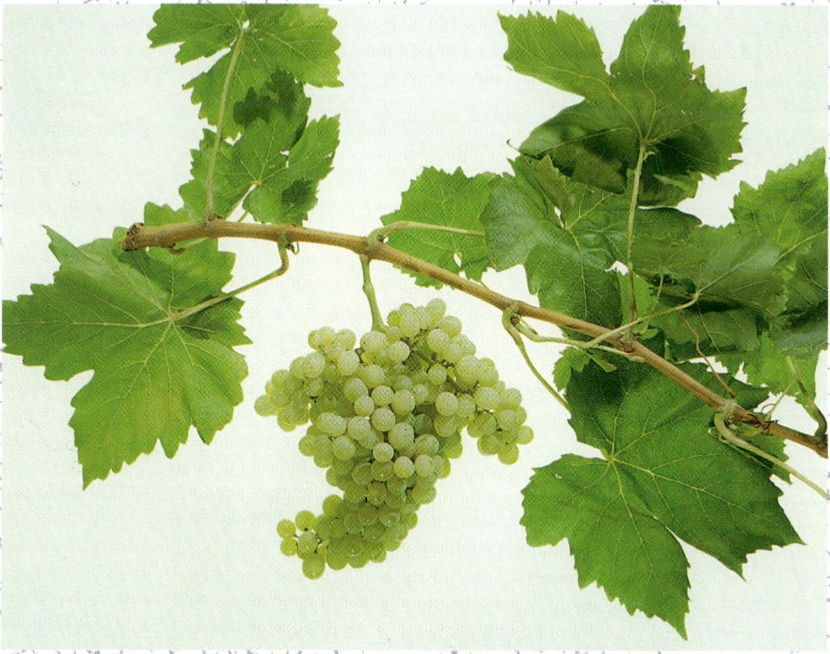

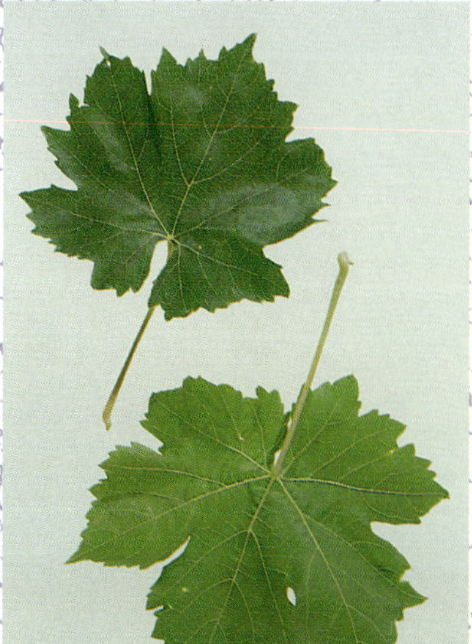

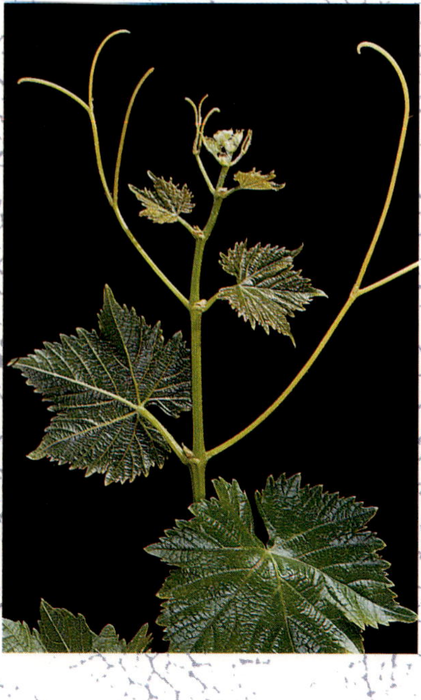

Sultana

Sultana is primarily a drying grape but in some seasons more sultanas are crushed for wine in Australia than grapes of any other single variety. There are about 16 000 ha of Sultana in Australia, almost all in the irrigation areas of the Murray Valley downstream from Swan Hill, and up to one-quarter of the crop may be used for wine. There are about 107 000 ha in California, where it is called Thompson Seedless to distinguish it from another variety introduced earlier erroneously as Sultana. An even larger proportion of the crop is used for wine in California and there are also districts which specialise in the production of table grapes. In Western Europe the variety is known as Sultanine or Sultanina. It seems to have originated in Asia Minor or the Middle East and is grown for dried fruit throughout the area from Greece to Afghanistan and north into the neighbouring republics. There are also small plantings in South Africa and South America.

Sultana is a fairly vigorous variety with a spreading habit of growth. It requires cane pruning because the buds at the base of the canes are rarely fruitful. It has medium, slightly 3-lobed leaves which are smooth and hairless and of a fresh green colour approached only by Grenache and Pedro Ximenez among the common wine varieties. Bunches are generally large and well filled, becoming smaller and looser with increasing crop, and the stalk is usually green and easily cut for hand harvesting. Berries are small, oval and seedless, with a firm flesh and a thin skin sensitive to damage by rain when ripe. Sultana is one of the few varieties seriously attacked by black spot and is more sensitive than most wine varieties to both downy and powdery mildew. It needs a clear, warm climate to promote the differentiation of fruitful buds but excessive heat can interfere with fruit set and berry development.

Sultana is more difficult to process for wine than specialised wine grapes because of its firm flesh and lack of seeds. However, it has a good acidity and produces fresh, rather neutral wines which are quite attractive on their own and form an excellent base for sparkling wines or for blending with highly flavoured varieties. The lack of seeds and the firm flesh make Sultana attractive as a table grape even at its natural berry size. The size can be increased by various cultural treatments to approach that of seeded table grapes but this is accompanied by loss of flavour.

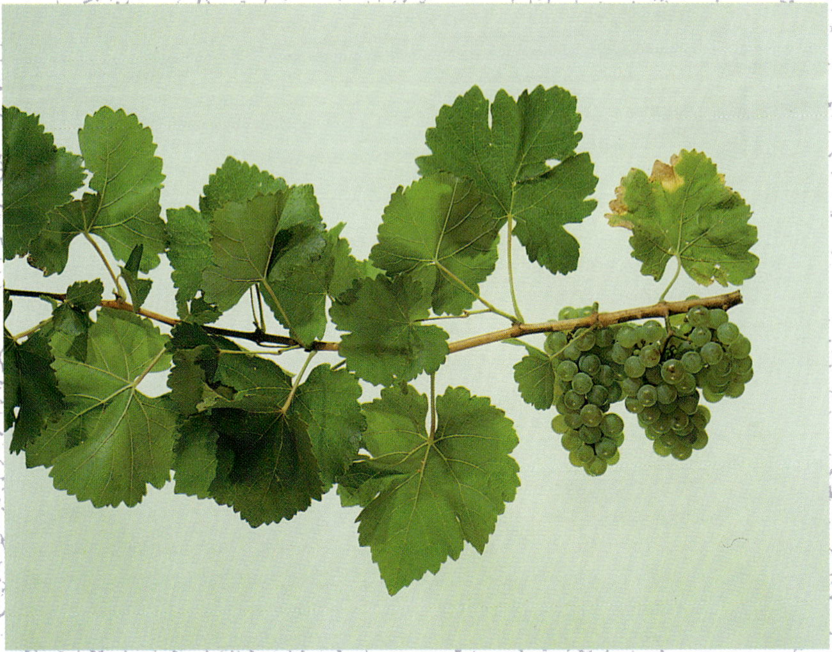

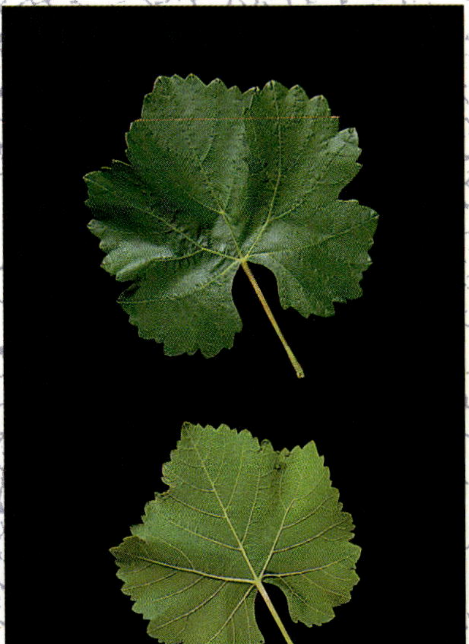

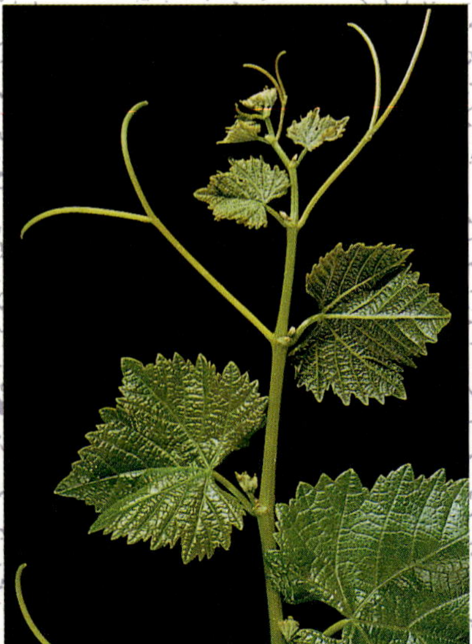

Sylvaner

Sylvaner, spelt Silvaner in Germany, was formerly the leading wine grape of that country, but in recent years it has declined considerably with the rise of newly bred varieties such as Kerner, Scheurebe and Bacchus. From 18 781 ha in 1964 the area of Sylvaner dropped to 7600 ha in 1994, while the combined area of the other three varieties rose from 349 ha to 15 000 ha. The variety may have originated in Austria, although there is only a small area there, and it can be found under various names through central and south-eastern Europe. In France the area has remained fairly steady and there were 2617 ha in 1988. There were 588 ha in California in 1982, but this had decreased to 75 ha by 1992. A few small plantings have been made in Australia.

Sylvaner is a vigorous, rather upright-growing and productive variety. It has clear green, slightly 3-lobed, smooth and undulating leaves without hair on the lower surface. The petiolar sinus is a narrow V. The bunches are small to medium in size, slightly conical and well filled with medium, round berries.

Under some circumstances Sylvaner can give a fruity, distinctive wine, but usually it gives a neutral wine which is well suited to blending. One of the newly bred German varieties, Morio Muskat, which is considered rather highly flavoured for use alone, has found a place as a blending partner for Sylvaner.

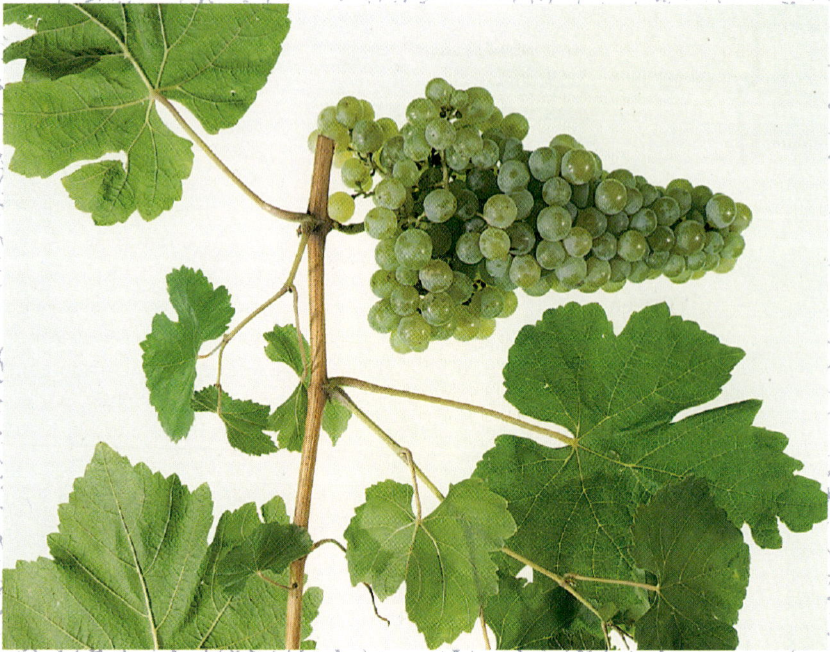

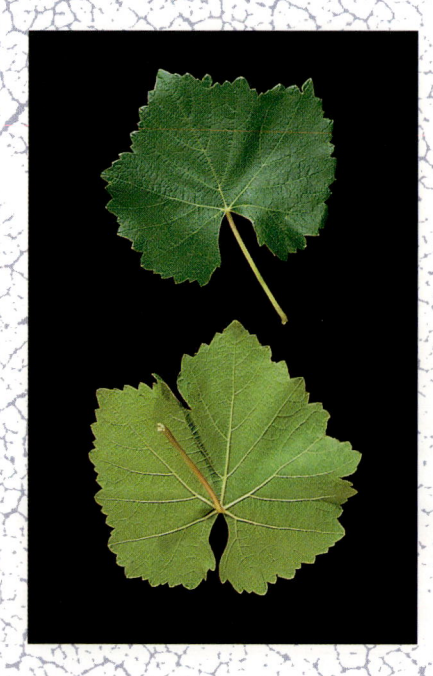

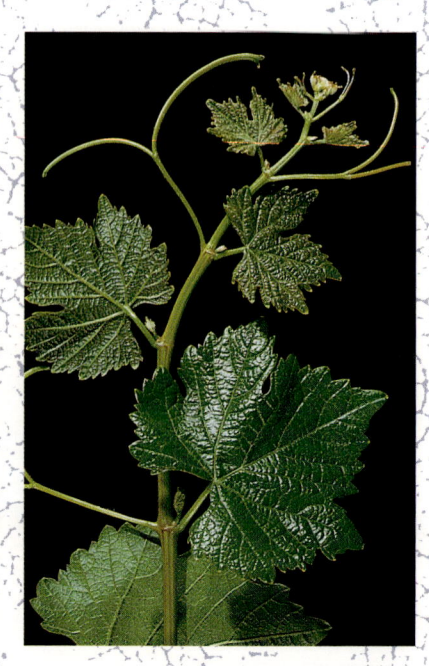

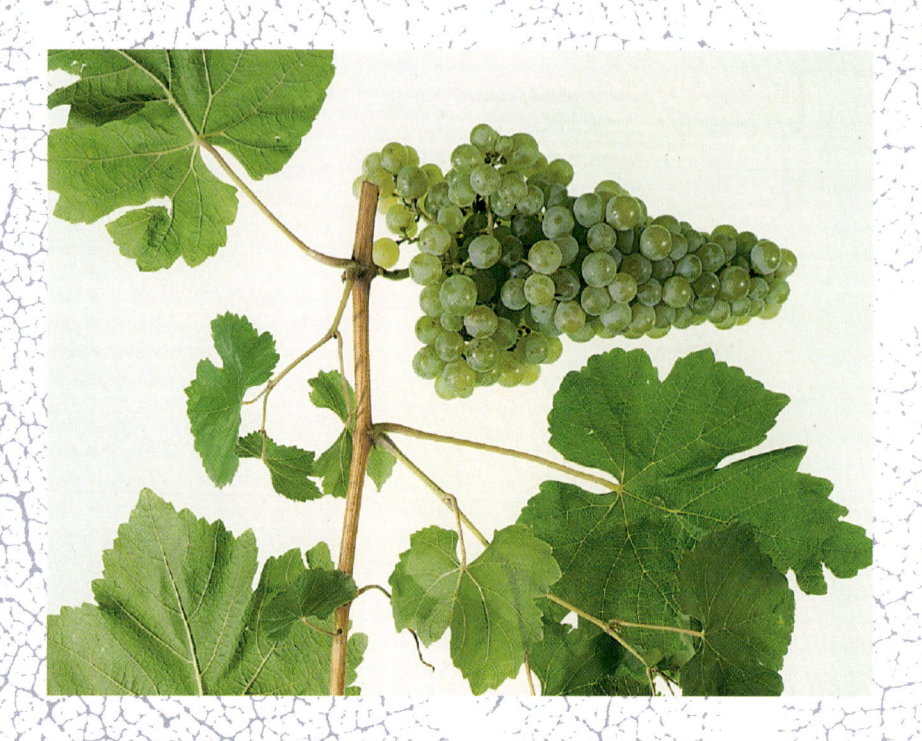

Taminga

Taminga is a grape variety bred by the CSIRO Division of Horticulture at its Merbein laboratory. It is a cross of the Merbein selection MH 29-56 (Planta Pedralba x Sultana) and Traminer. The cross was made in 1970 and the variety released in 1982. Approximately 200 ha have been planted in Australia, mainly in the Murray River Valley.

Taminga is a high yielding, late ripening variety which produces high quality wines with a distinctive aromatic bouquet. The leaves are medium in size, orbicular, dark green with a glossy upper surface, slightly involute, rough when young but becoming almost smooth when mature. They are glabrous above and have sparse, short, bristly hairs below. They are slightly to moderately 3-lobed, the petiolar sinus is generally a narrow V, teeth are broad and rounded, alternating large and small. The bunches are small to medium, conical, winged, well filled to compact, the flowers are hermaphrodite, berries medium, round, white with a distinctive varietal character. The seeds are medium in size and hard. The canes are finely ribbed, light brown in colour tending to dark brown at the nodes and sometimes on the ribs, internodes are short. The variety ripens late February to early March in the Murray Valley.

Mean yields have been around 40 tonnes per hectare. The juice composition has been good with low pH and high titratable acidity. It has been used commercially in Australia to produce a high quality botrytised sweet wine, and for blending with more neutral flavoured varieties to improve their flavour.

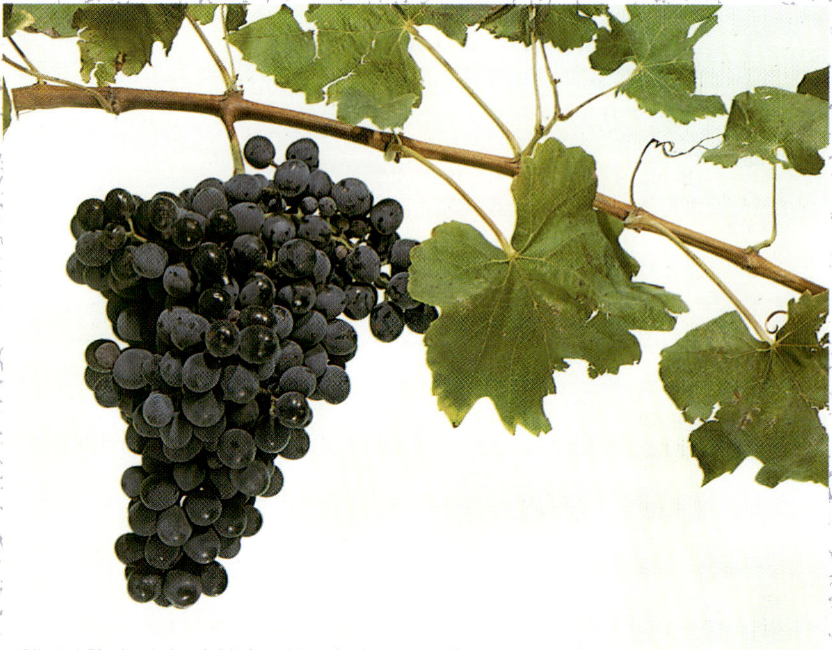

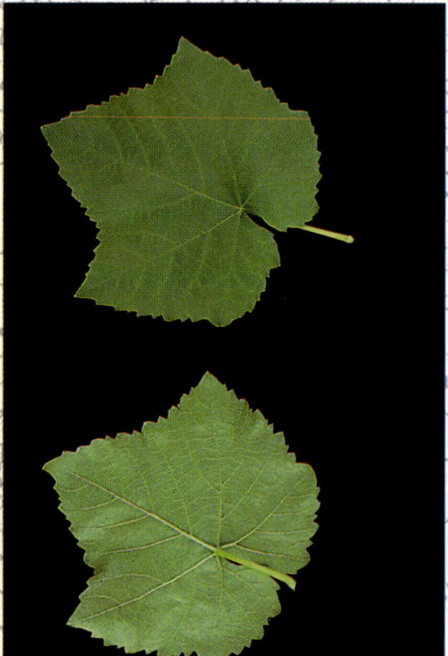

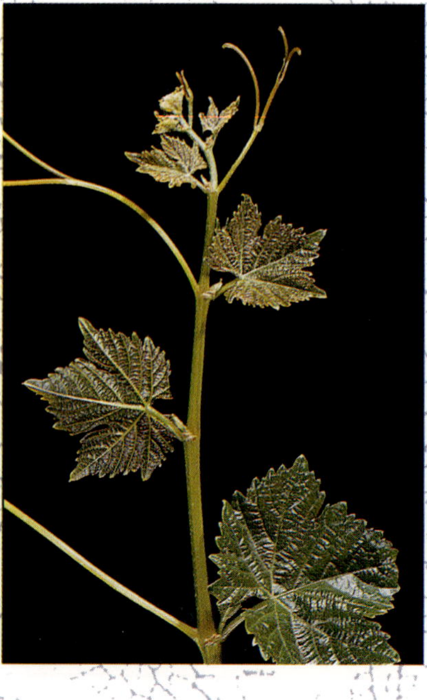

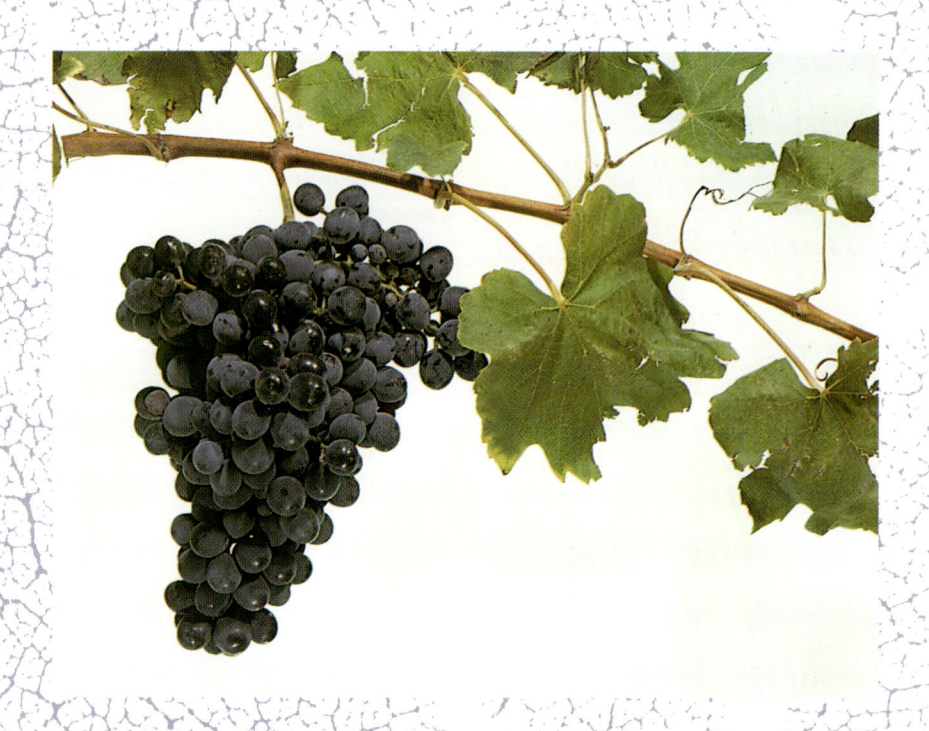

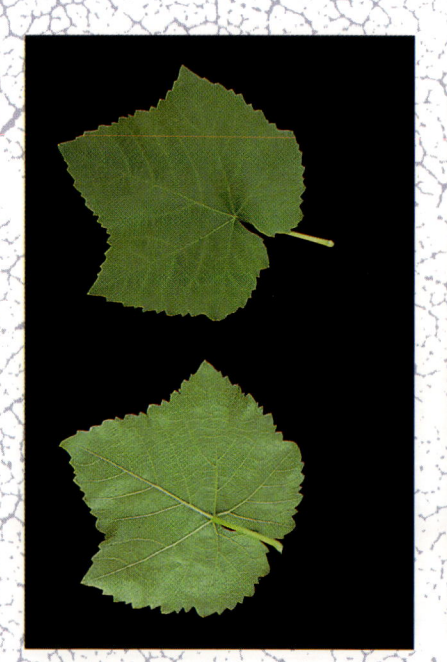

Tarrango

Tarrango is an Australian variety bred by the CSIRO Division of Horticulture at its Merbein laboratory. It is a cross of the Portuguese port variety Touriga and the ubiquitous multi-purpose variety Sultana. Approximately 200 ha of Tarrango have been planted in Australia, mainly in the warmer regions along the Murray River.

Tarrango is a vigorous variety which, like Sultana, requires either cane or minimal pruning because of the low fruitfulness of the basal buds. It has medium sized, orbicular, light green leaves which are plane, smooth, glabrous and entire to 3- or 5-lobed. The petiolar sinus is closed with overlapping edges. The teeth are broad and shallow. Bunches are large, conical and well filled. Flowers are hermaphrodite. The berries are medium sized, black and short oval in shape with small hard seeds. Tarrango ripens late, about the first week in April in the warm, irrigated regions of Australia. The juice has a low pH and high titratable acidity. Mean yields have been in the order of 32 tonnes per hectare. The wines have a distinctive fruity flavour and bright hue. They have a high tartaric acid/malic acid ratio which combined with low pH gives excellent colour stability and hue. The wine has also been used as a successful sparkling wine base.

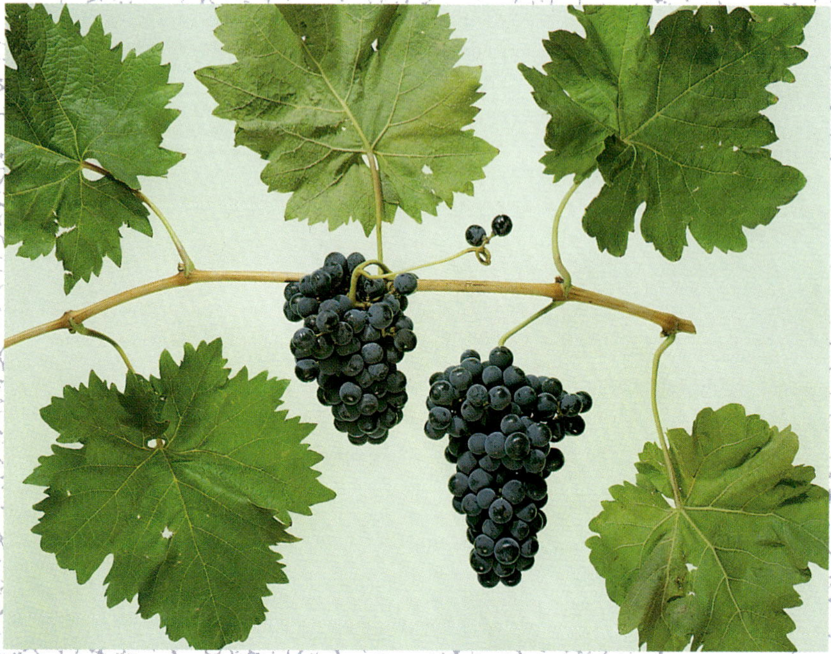

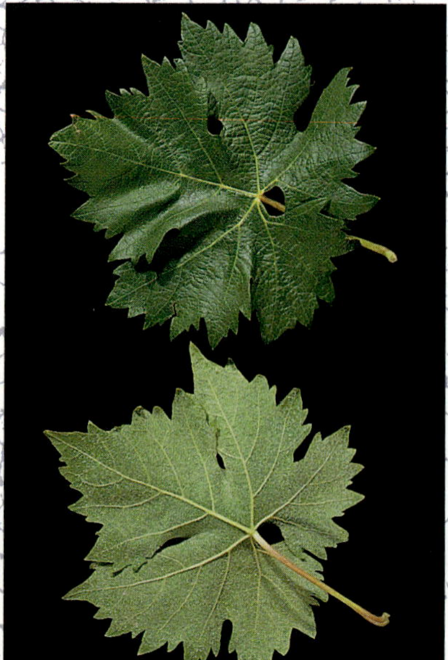

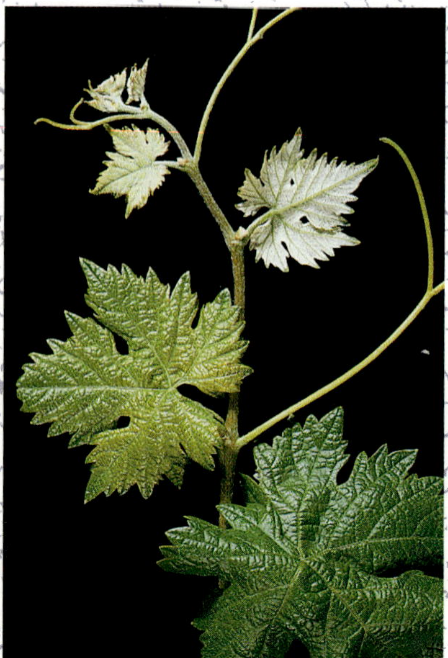

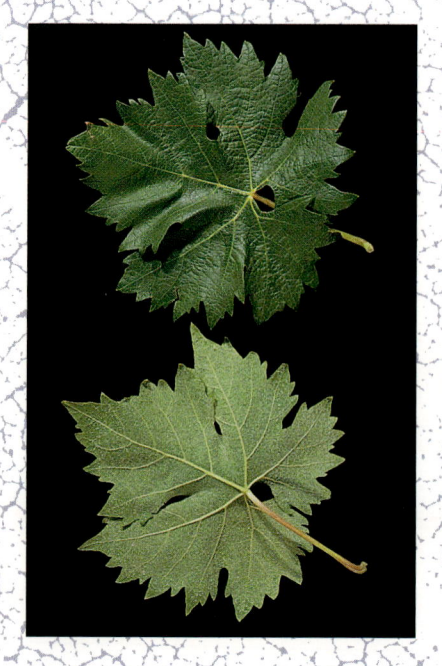

Tempranillo

In Spain, Tempranillo is one of the most highly regarded of the varieties used for making red wine and about 40 000 ha have been planted. Blended with Carignan it makes the best wines of the Rioja. In La Mancha it is known as Tinta Fino or Cencibel and is used in the claretes of Valdepenas and Manzanares. In Portugal it is known as Tinta Roriz, and Negretto in Italy may perhaps be the same variety. Tempranillo was authorised in 1976 for planting in the Mediterranean region of France and by 1988 there were 2314 ha. It is an important variety in Argentina, where there were 12 000 ha in 1986. In California, where it is called Valdepenas, there are about 220 ha. Only small experimental plantings have been made in Australia.

Tempranillo is a vigorous variety with an upright habit of growth. At Merbein, the clone imported from California is noticeably more vigorous than the clone imported from Spain. The variety has large, thick, 5-lobed leaves with tufted hairs on the lower surface and a closed petiolar sinus with overlapping edges. The leaves often have a pale, limp appearance as they expand to full size. The bunches are large and well shouldered, well filled to compact, with berries which are quite variable in size and shape and which have a heavy bloom.

Tempranillo gives table wines with a good colour intensity and a slightly bluish hue. The wines mature quickly and are ready for bottling and drinking in the year of vintage. In Portugal, as Tinta Roriz, the variety is ranked in the highest quality group for producing port.

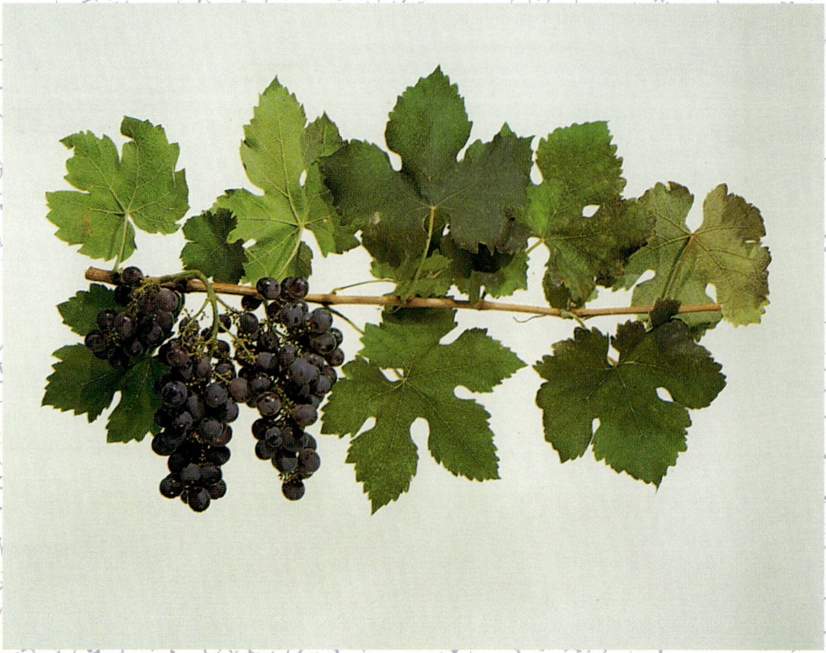

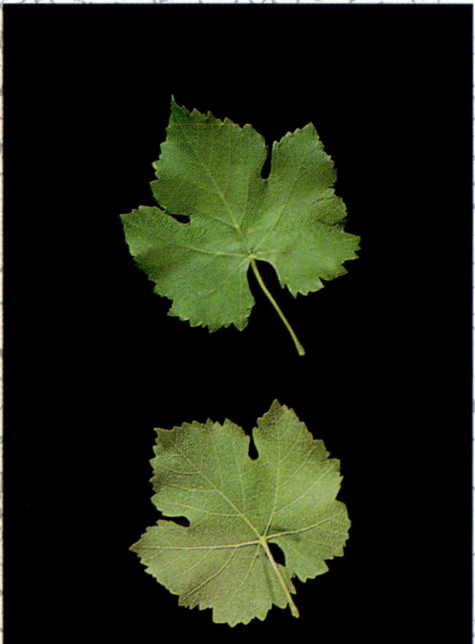

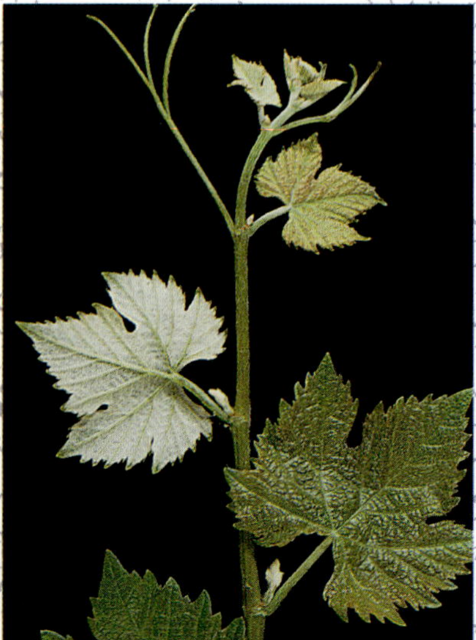

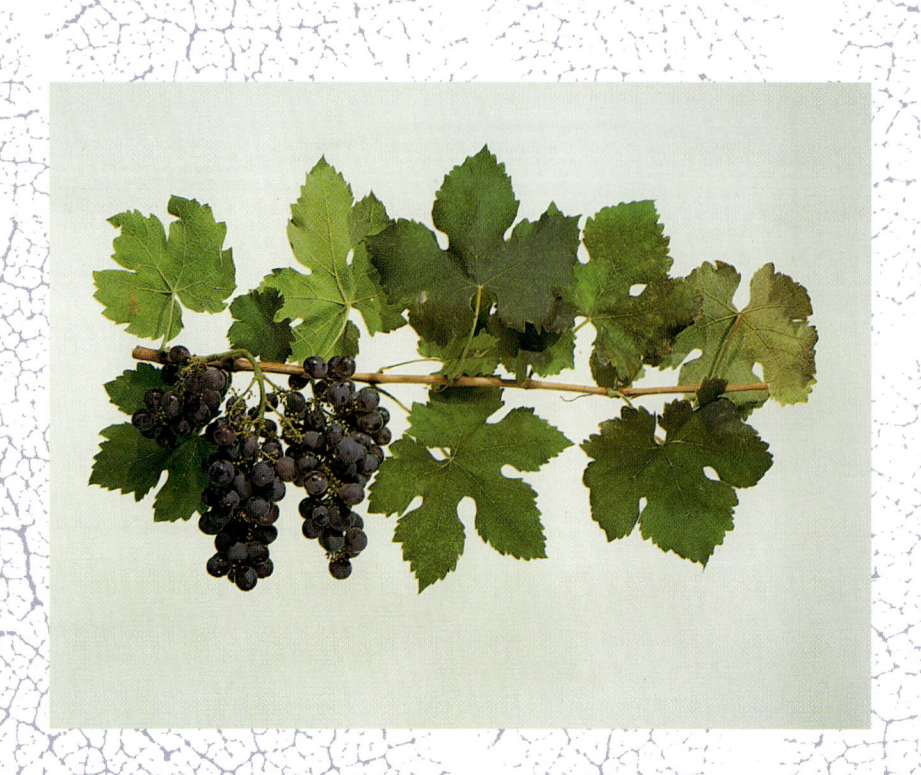

Terret Noir

The Terrets are recommended varieties from the Languedoc area in the south of France. Three colour variants are known — white, grey and black — and chimaeras are sometimes seen in all combinations. There are about 3000 ha each of Terret Blanc and Terret Gris, although the area is decreasing, but only about 800 ha of Terret Noir. Terret Gris is also known as Terret Bourret or sometimes just Bourret. In Australia there is a small area of Terret Noir in the Barossa Valley.

Terret Noir shows a late budburst and is a fairly vigorous variety with upright growth. Leaves are medium in size, 3- to 5-lobed, smooth and hairless above, and with tufted hairs on the lower surface. Bunches are medium, often with a woody stalk, often winged, potentially well filled but sometimes straggly because of poor setting. Berries are medium, short ellipsoid, not deeply coloured, with a firm skin resistant to bunch rot.

Terret Noir is listed among the varieties to be used for a number of French appellation wines, including Chateauneuf-du-Pape and Cotes du Rhone, but it is unlikely that it would ever be more than a minor component of the wine. Terret Noir alone gives wines without much colour but which are light, fresh and distinctive, capable of combining well with those from more full bodied varieties such as Mataro.

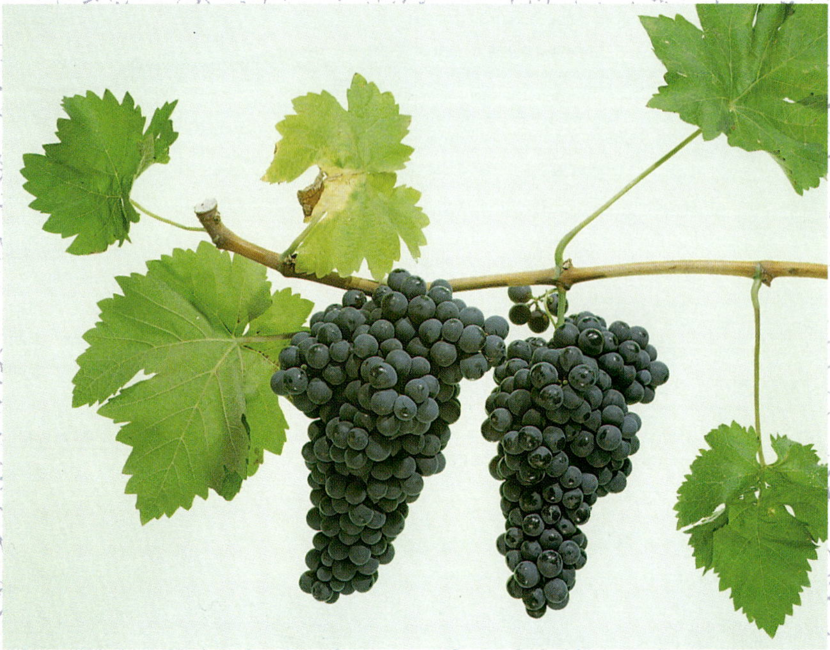

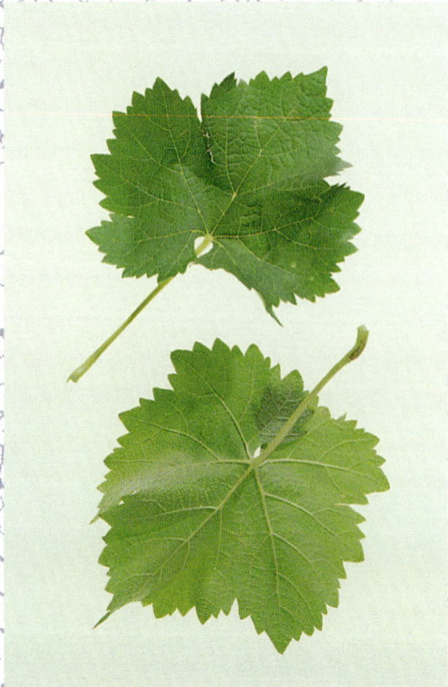

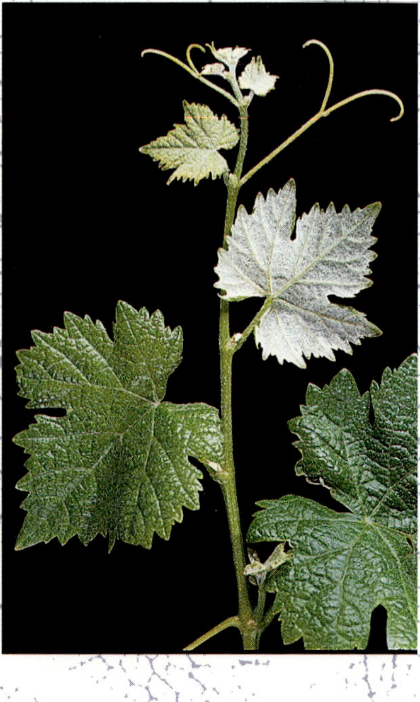

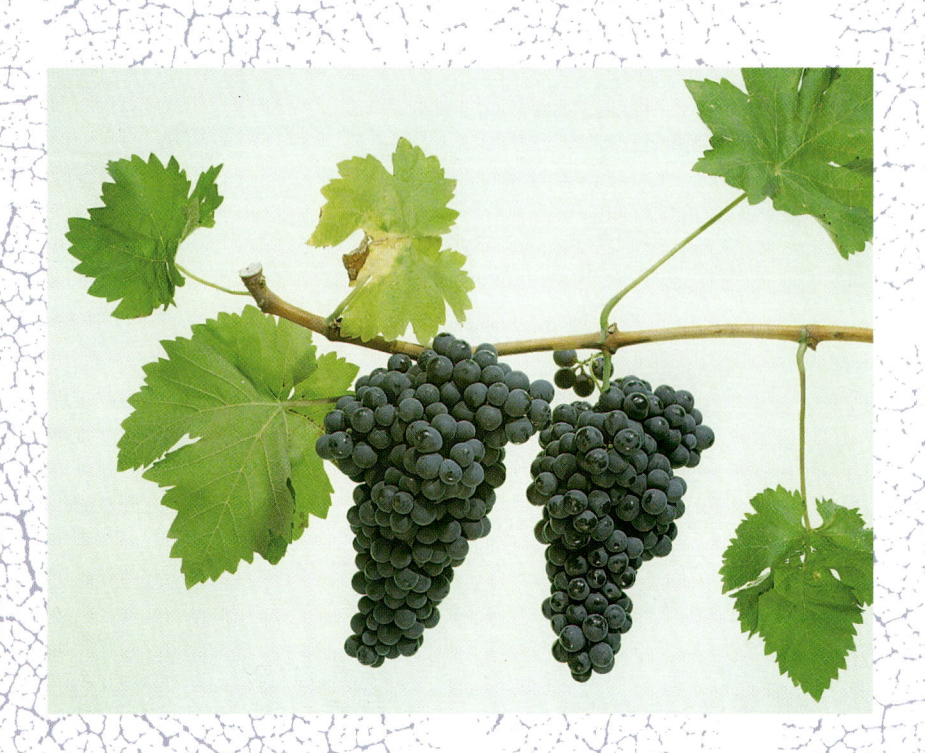

Tinta Amarella

Tinta Amarella is best known as a port variety. It is widely grown in the Douro Valley, where it is regarded as a good, rather than a very good, variety. It seems to be little grown outside Portugal, but it is distinctly different from the other common port varieties and it is thought that it may have come from France. It could perhaps have been the Amarot of Landes, listed among Busby's imports into Australia. There are about 30 ha in Australia, all in South Australia, where the variety is known as Portugal. It is not uncommon as odd vines in plantings of other varieties and this seems to have led to confusion with Malbec: some intended plantings of Malbec have been actually made with Tinta Amarella.

Tinta Amarella is a vigorous variety with a spreading habit of growth. It has medium, rough and undulating, 3- to 5-lobed leaves with usually a distinct overlap at the petiolar sinus. The leaves are clear green in colour with sometimes a yellow tinge, hairless above with fairly thick tufted hairs below. Bunches are medium, conical and compact, with short stalks. Berries are medium, nearly round when not pressed out of shape, with a moderate bloom and fairly thick skin.

Wines from Tinta Amarella have a reasonable colour but no particular varietal or other special character. They could perhaps be used in blending to soften harsh or astringent wines.

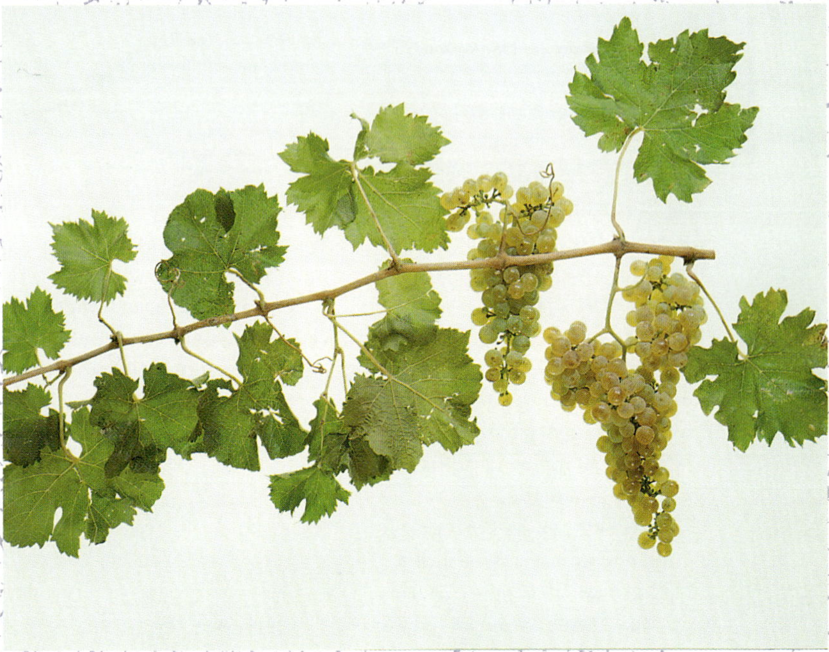

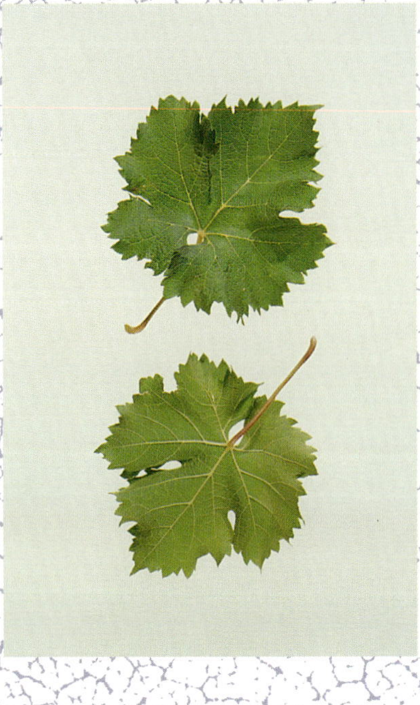

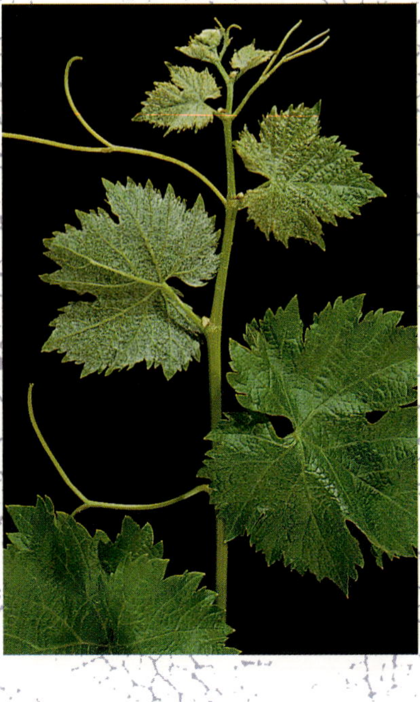

Tocai Friulano

This variety probably originated in France, where it is known as Sauvignonasse, and appears to be related to Sauvignon Blanc. However it would seem more appropriate to use the Italian name, Tocai Friulano, because the variety has almost disappeared from France while it has become very popular in the Friuli-Venezia Giulia region of Italy, where there would be several thousand hectares planted. Tocai Friulano is also the preferred name in Argentina where there are 1500 ha. In Chile, where there are more than 3000 ha, it has been erroneously called Sauvignon. It is widespread in Australia, but only in plantings of other varieties, having been found in Mudgee, the Goulburn Valley and Great Western.

Tocai Friulano is a vigorous variety with rather upright growth. It has leaves which are medium or above in size, more or less 3-lobed, clear light green in colour, practically hairless on the lower surface, and tend to fold inwards with very frilled edges. The petiolar sinus is often closed with overlapping edges. Bunches are small to medium, well filled, cylindrical and sometimes winged, with small to medium, round, juicy berries. The buds at the base of the canes are not very fruitful and long pruning is needed for satisfactory cropping. A system of fully overhead trellis with hanging canes is popular in Italy.

Wines of Tocai Friulano have a delicate but definite bouquet and a slight bitterness on the palate which is appreciated in Italy. The proportion of Tocai Friulano in some Australian plantings may be high enough to influence the character of the wine produced.

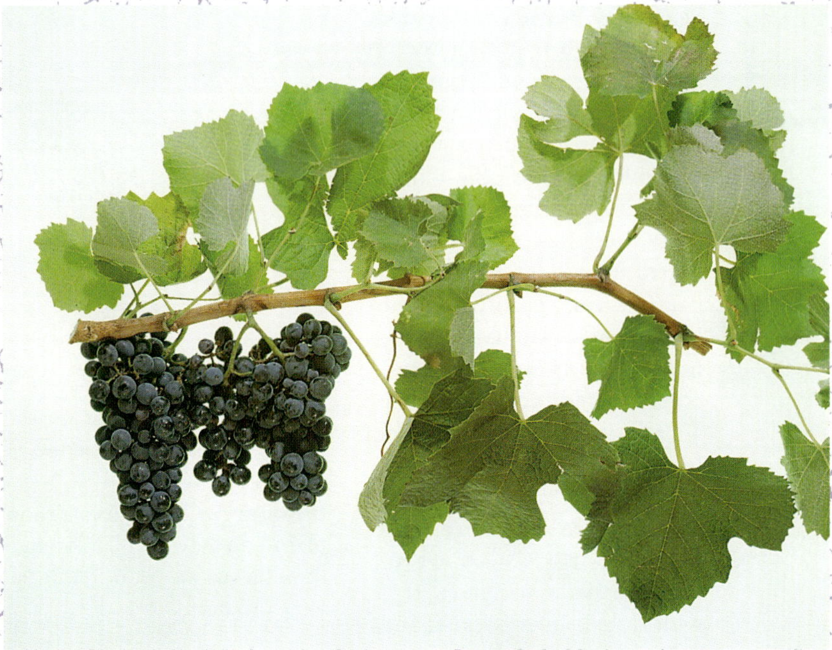

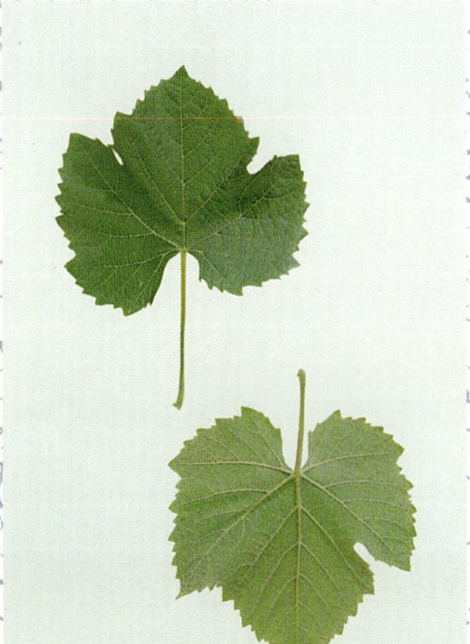

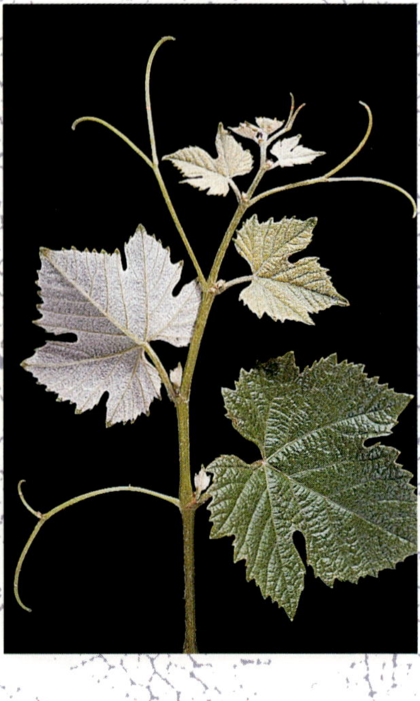

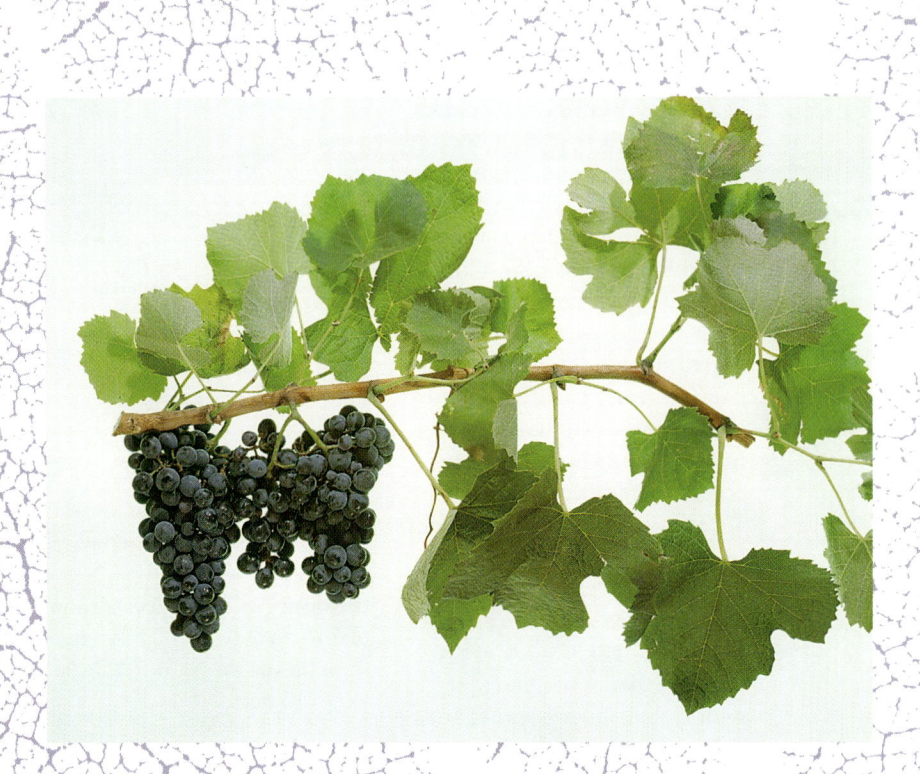

Touriga

Touriga is the most widely planted of the top quality group of the port varieties in the Douro Valley in Portugal. It is also grown in other parts of Portugal for other types of wine, but there seems to be very little outside Portugal. There are a few hectares in California and it is probably one of the 'port sorts' in South Africa. There were 39 ha of Touriga in Australia by 1990 with 28 ha of this being in South Australia.

Touriga is a fairly vigorous variety with a trailing habit of growth. It has proved to be one of the varieties least susceptible to salt burn from overhead irrigation. It has medium, mid green leaves which tend to fold in about the midrib and can vary from almost entire to distinctly 5-lobed. The upper surface is smooth, sometimes with the veins depressed, and may have a few cobwebby hairs. The lower surface has tufted cobwebby hairs. The petiolar sinus is V-shaped, often widely open. The bunches are small to medium, more or less conical, with rather small, short oval berries with a heavy bloom.

The pre-eminence of Touriga as a port variety is well justified. It produces fortified wines of excellent colour and flavour in Australia also.

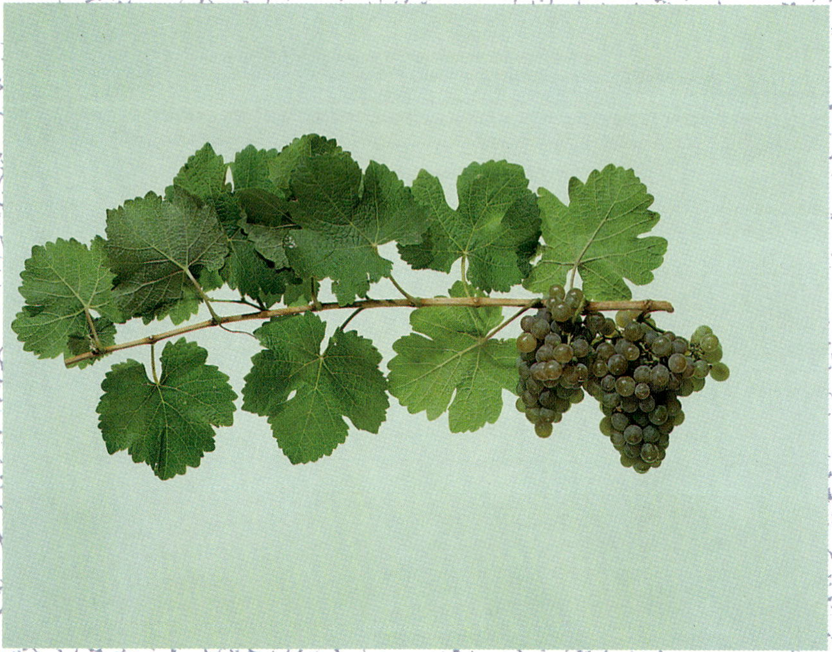

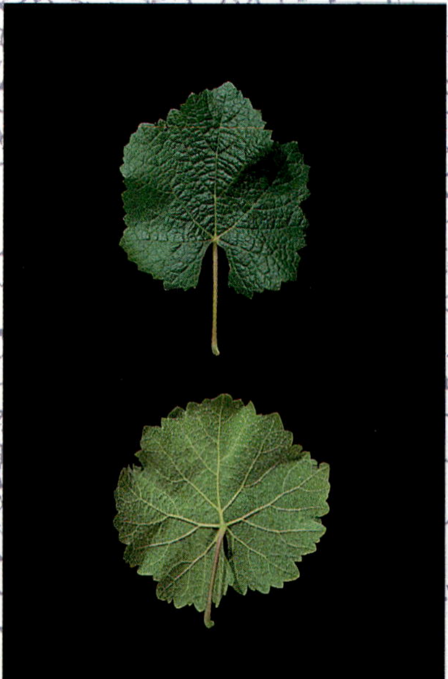

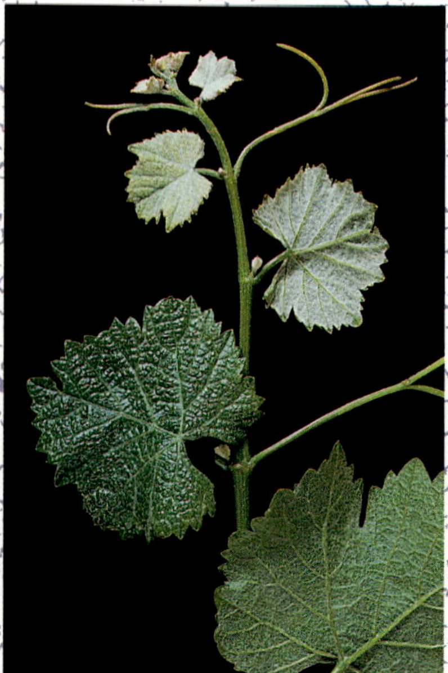

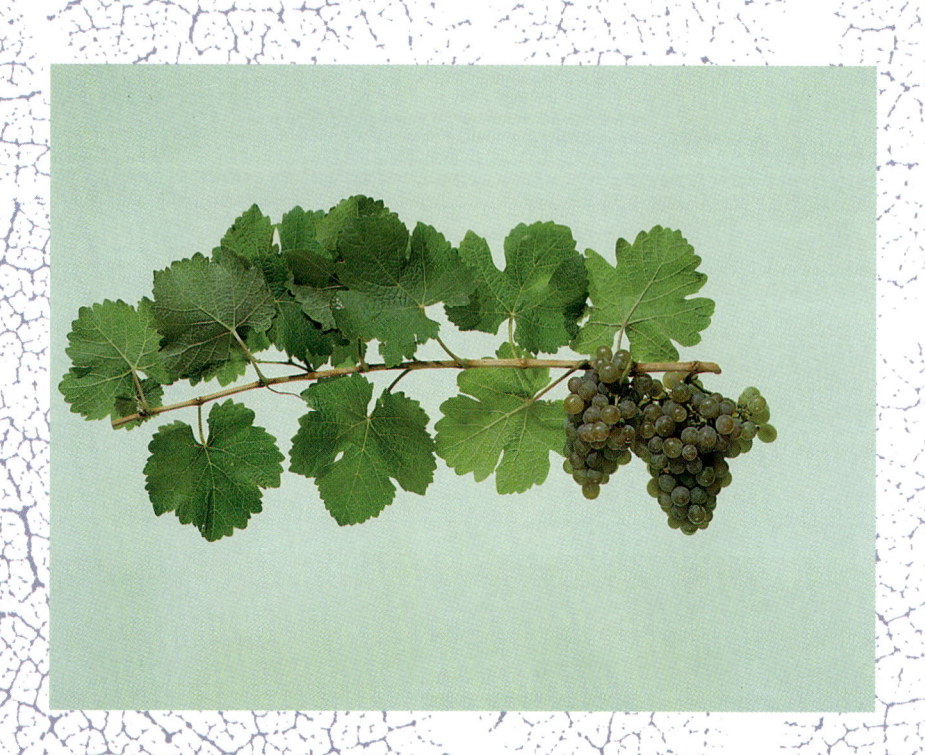

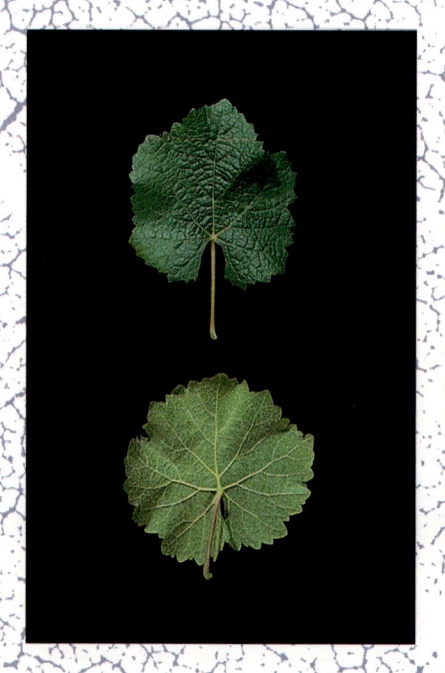

Traminer

Traminer is another old variety showing primitive characters. It is valued mainly for the high aromatic character of its wine and as there are large clonal differences in this respect the more aromatic clones are sometimes distinguished as Gewurztraminer. Being a specialised variety it is not required in large quantities and thus there are about 775 ha in Germany, 700 ha in Austria, 2590 ha in France and 670 ha in California. Some people in Germany and central Europe prefer the wines to be not quite as aromatic and some of the plantings are of this type. The variety has been in Australia for a long time but it is only recently that much has been planted with the area reaching about 800 ha by 1983, although this had decreased to 548 ha by 1992.

Traminer is a vigorous variety in cool climates, but where the climate is hot it is less vigorous and slow to reach its full development. In cool climates the leaves are large, slightly 3-lobed, very rough, and downy on the lower surface, and have the petiolar sinus closed with the edges overlapping. In hotter areas the leaves tend to be smaller and smoother with the petiolar sinus not quite closed. The bunches are small and tight with very short bunch stalks and borne at the lower buds of closely budded shoots, making it a very difficult variety for hand harvesting. The berries are small and oval, sometimes showing hen and chicken in cool areas, and are an increasingly intense pink in colour the cooler the climate.

Wines from Traminer grown in hot climates are likely to be undistinguished although this is not always the case. Under cooler conditions, and particularly when the berries are infected with noble rot, it can give luscious wines with an intense spicy varietal character.

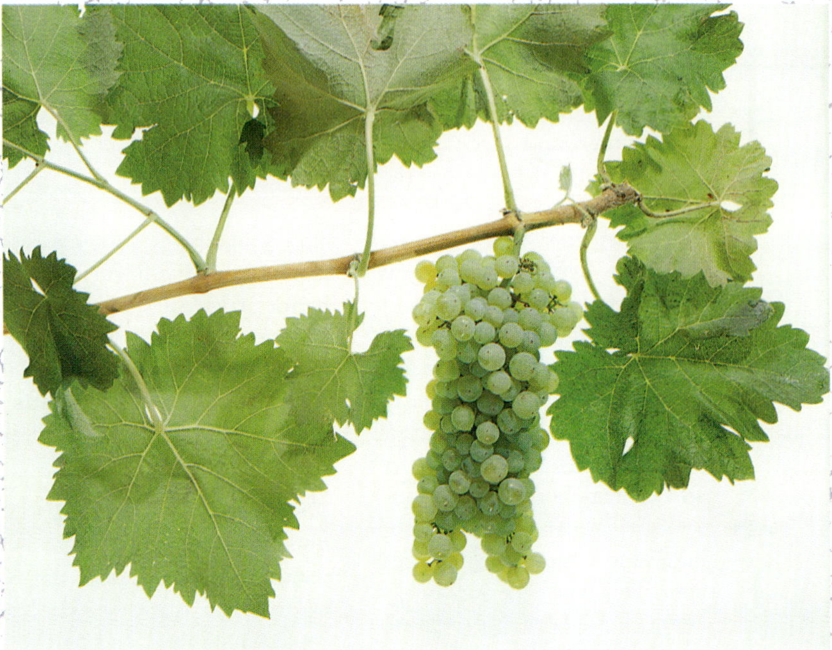

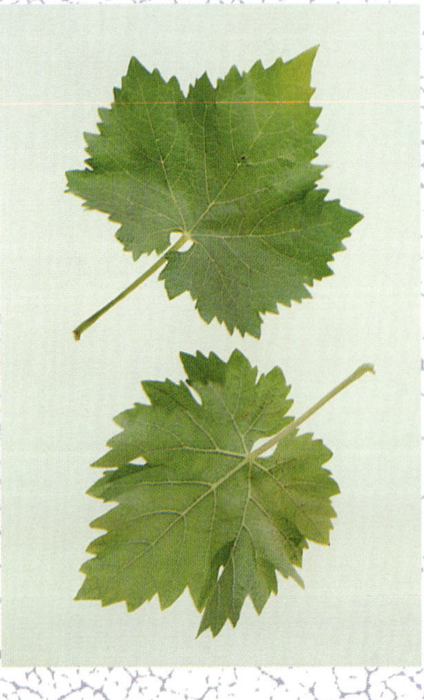

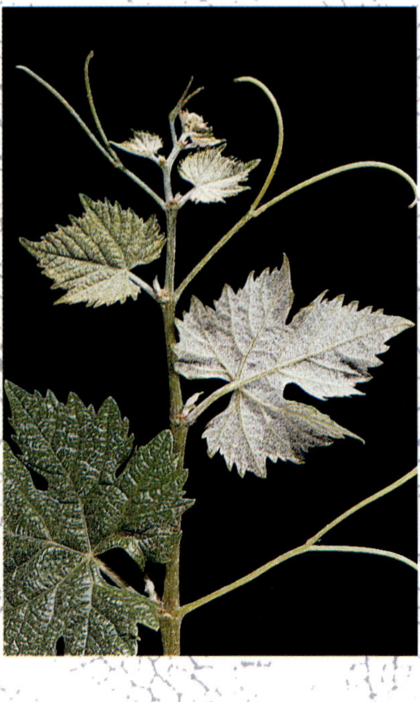

Trebbiano

Trebbiano is the Italian name of this variety and it is more specifically known as Trebbiano Toscano to distinguish it from several other similar varieties also known as Trebbiano. It is by far the leading white wine grape in Italy with plantings of the order of 50 000 ha, about three times the area of the next most important white variety, Trebbiano Romagnolo. It spread into the south of France and from there into the Cognac area when a new variety was needed to take the place of Folle Blanche which suffers from bunch rot when grafted. In 1958 there were 62 000 ha planted in France. This had increased to 128 000 ha by 1979 but then decreased to 103 000 ha by 1988. The official French name is Ugni Blanc but in the Cognac area it is known as Saint Emilion. Elsewhere it is a minor variety, with about 1000 ha in South Africa, 240 ha in California and 2200 ha in Argentina. In Australia there are about 1000 ha, with just over half in New South Wales and most of the rest in South Australia. It has sometimes been known as White Shiraz or White Hermitage, but it does not appear to be in any way related to Shiraz and is not grown in the Hermitage vineyards in France.

Trebbiano is a vigorous and productive variety which bursts late and has rather upright growth. The leaves are large, thick, rough and undulating, mostly slightly 3-lobed and tending to fold inwards, dark green above and hairy below. The bunches are well filled, basically long and cylindrical but with a large wing and often with a very characteristic forked tip. The berries are medium and round, ripen rather late becoming golden where exposed. They have a tough skin but soft flesh and juice badly when mechanically harvested. The variety has proved very susceptible to root-knot nematode on the coastal sands of southern France where, because phylloxera is not a problem, vinifera vines are grown on their own roots.

In a cool climate Trebbiano gives acid wines which are excellent for distillation for brandy. In warmer areas it gives fresh neutral wines which blend well with more fruity types. It is an important component of a number of named wines in both France and Italy including the red wines of Chianti and Cotes du Rhone.

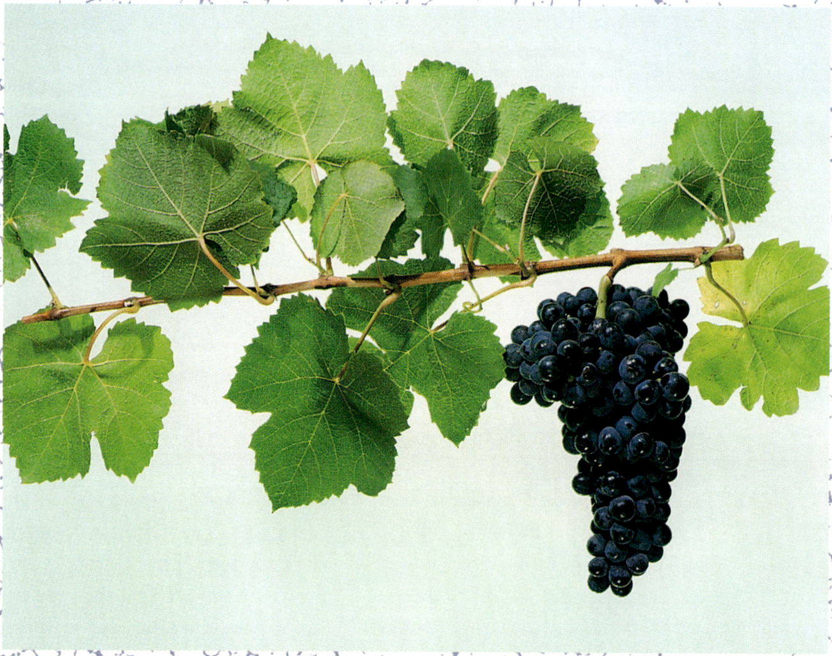

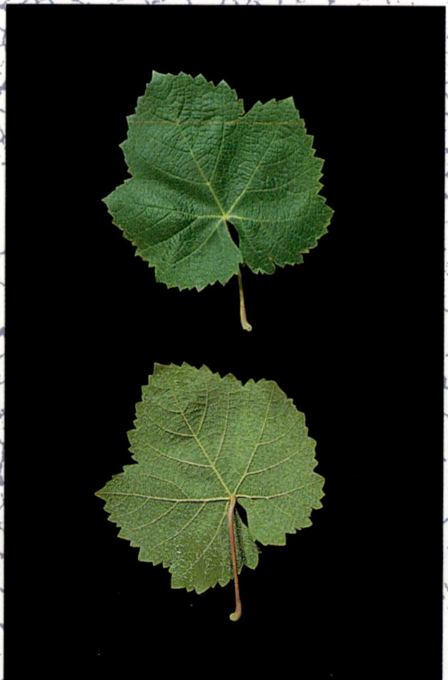

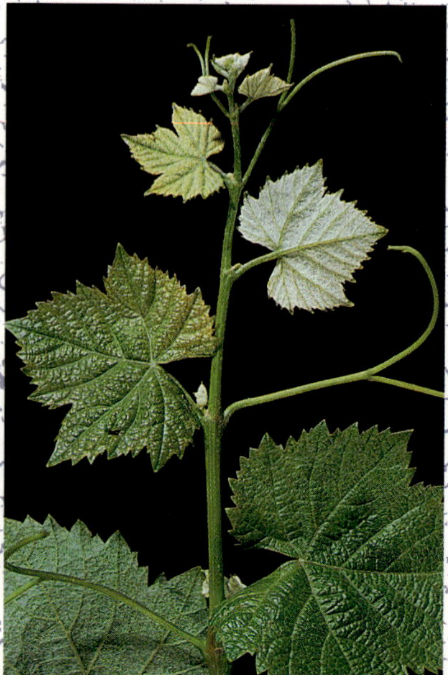

Valdiguie

Valdiguie was introduced into Australia from California as Napa Gamay. It is a variety which is authorised, not recommended, for the south of France and is grown mainly in Languedoc and Provence. As would be expected the area in France has declined, from 4900 ha in 1958 to 274 ha in 1988. Plantings in California reached 2476 ha by 1977 but had declined to 548 ha by 1992.

Valdiguie is a fairly vigorous and very productive variety. The leaves are large and rough, more or less definitely 3-lobed, tending to roll back at the edges and quite hairy on the lower surface. The petiolar sinus is a narrow U which is sometimes closed but not often overlapping. The bunches are conical, large and compact, with medium, slightly oval berries. In France the heavy yields are accompanied by poor sugar accumulation, but in warmer climates it seems to be possible to achieve good sugar concentration along with good acidity. The variety shows a modest degree of resistance to oidium.

In France, Valdiguie gives wines which are regarded as common and uninteresting. They have good colour but are lacking in alcohol and flavour. However, in California it is esteemed for the production of both red and rosé wines. Wines made at Merbein have scored well in blind tastings.

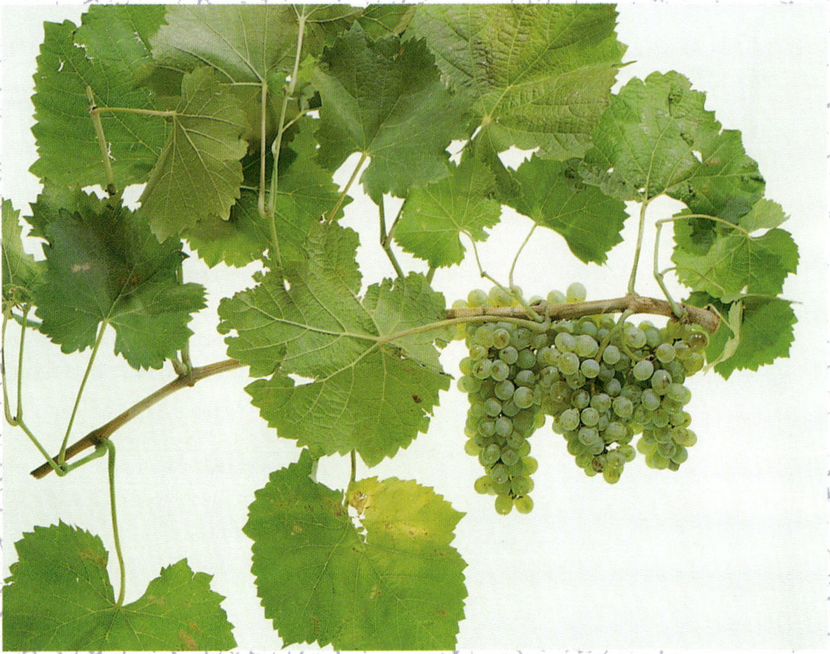

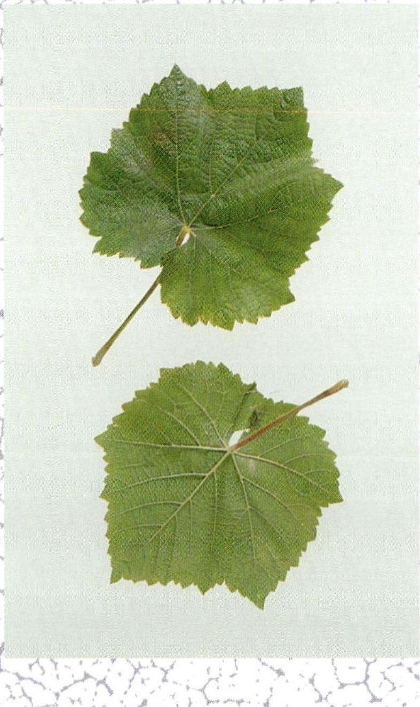

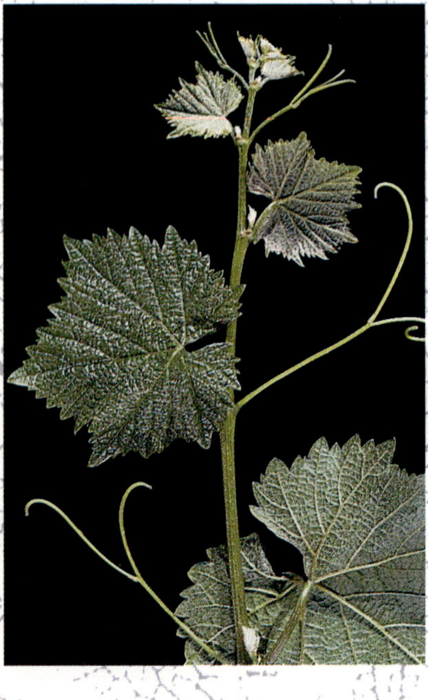

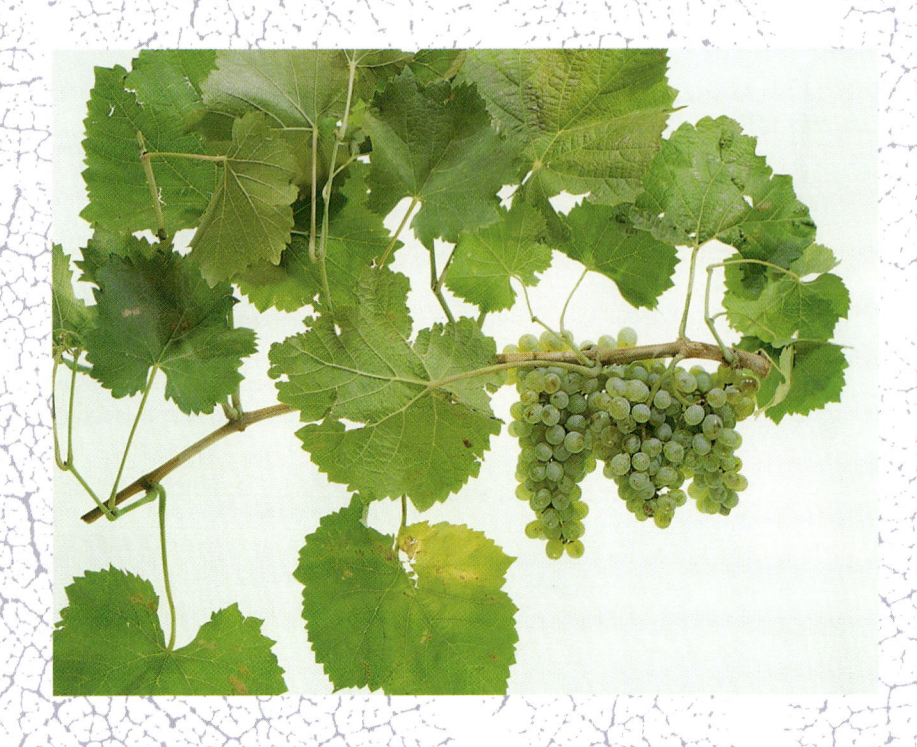

Verdelho

Verdelho is a Portuguese variety grown both on the island of Madeira and in the Douro Valley, where it is known as Gouveio. There are also some small plantings in the Loire Valley in France. In Australia there are 226 ha divided almost equally between Western Australia, South Australia and New South Wales. The name Madeira has sometimes been used for the variety in New South Wales, a possible source of confusion because the same name was erroneously used for some of the Semillon in South Australia.

Verdelho is a moderately vigorous and productive variety with a spreading habit of growth. It has medium, dull leaves, entire or slightly 3-lobed, with small tufts of hair on the veins and veinlets on the lower surface. The bunches are small, winged and loose with small, oval, greenish yellow berries becoming gold where exposed.

In Portugal Verdelho is used for fortified wines, in white port and as the predominant variety in one of the four types of madeira. In France it is used for dry white table wine. In Australia it is generally used for table wines and shows a strong and attractive varietal character.

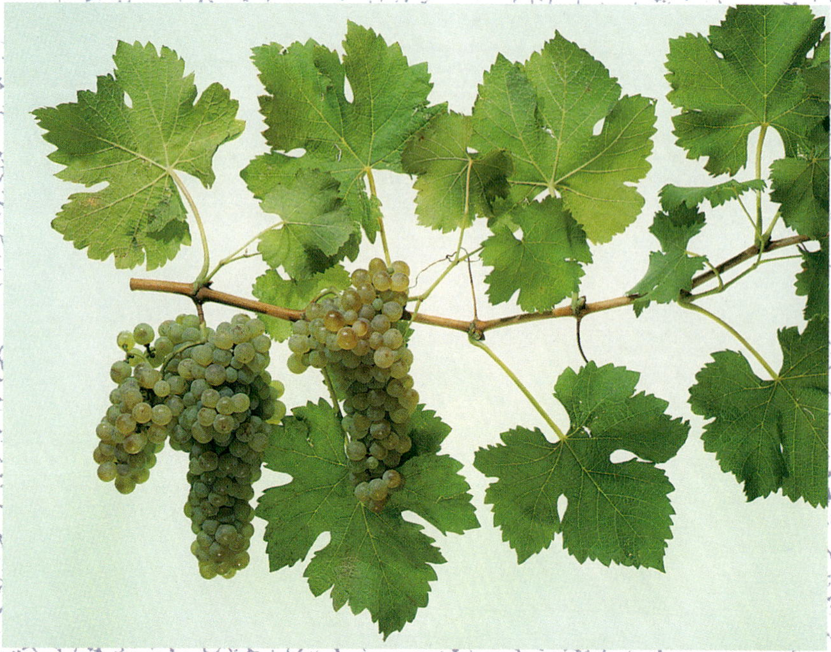

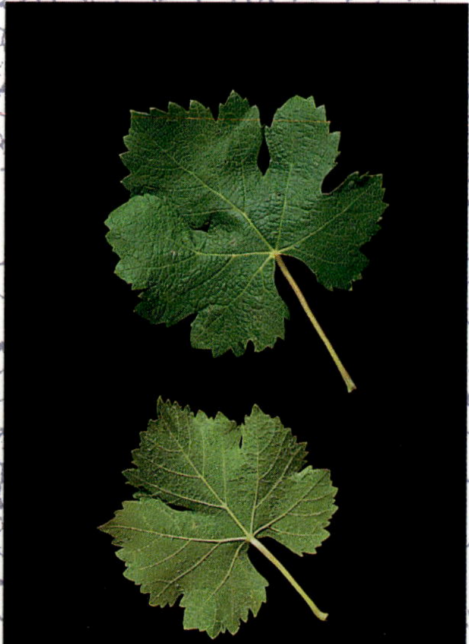

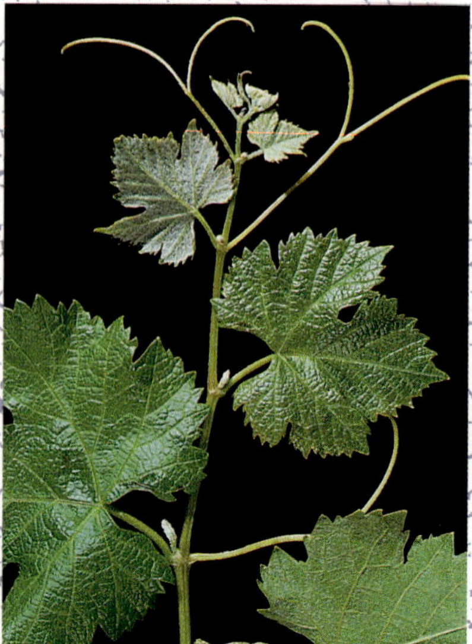

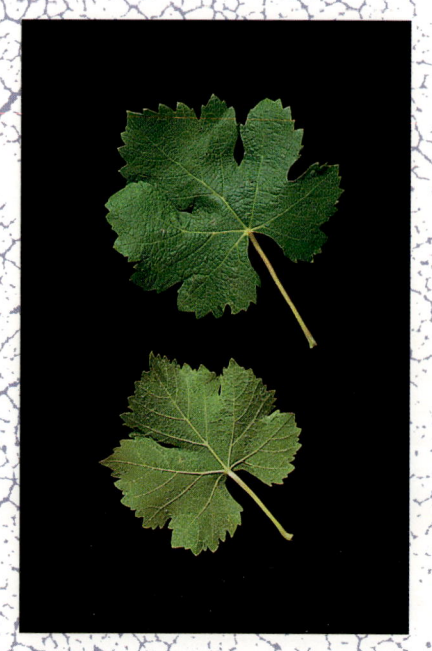

Viognier

Viognier is an old variety restricted to the right bank of the Rhone river south of Vienne in France. In the appellation Cote-Rotie it is blended with Shiraz to add perfume to the red wine, but in Condrieu it is vintaged alone to make a dry white wine with a floral scent and a long spicy aftertaste. Chateau Grillet, at the centre of Condrieu, with its 1.6 ha of Viognier, is the smallest vineyard with its own appellation in France and one of the most famous.

The vines have medium vigour and yields. The leaves are medium sized, orbicular, bullate and the petiolar sinus is an open V. The teeth are pointed and narrow. The leaves are downy below. The shoots are striated, clear green turning light brown in the sun, the nodes are pinkish and the tendrils are long and fleshy. The bunches are medium in size, conical, winged and compact. The berries are small, slightly oval, white with a thick skin and a light musky flavour.

Viognier has not been grown extensively in Australia but there has been a recent increase in interest in this variety. The variety is not very fruitful and should be cane pruned. The special aromatic character of the wine could make it useful as a specialty line in some of the cooler regions of Australia.

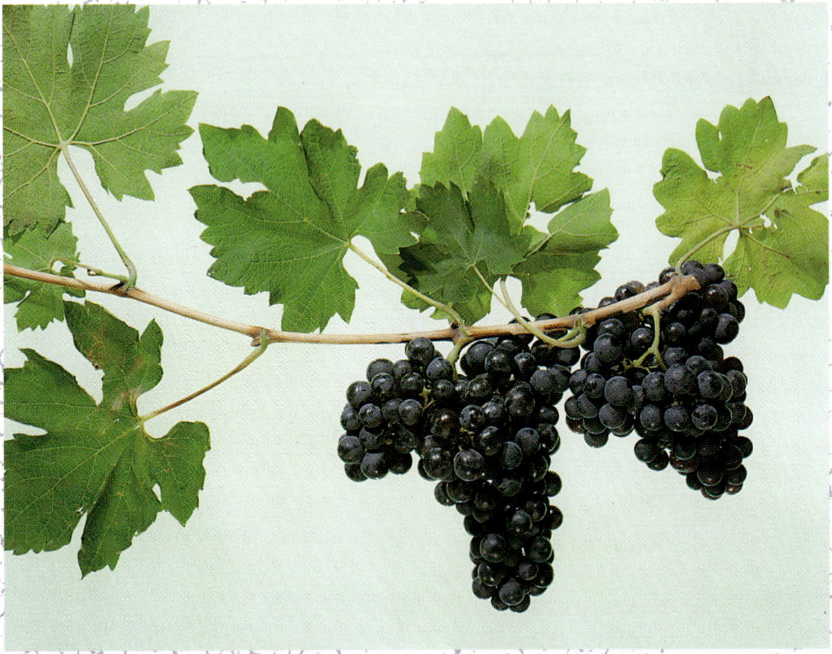

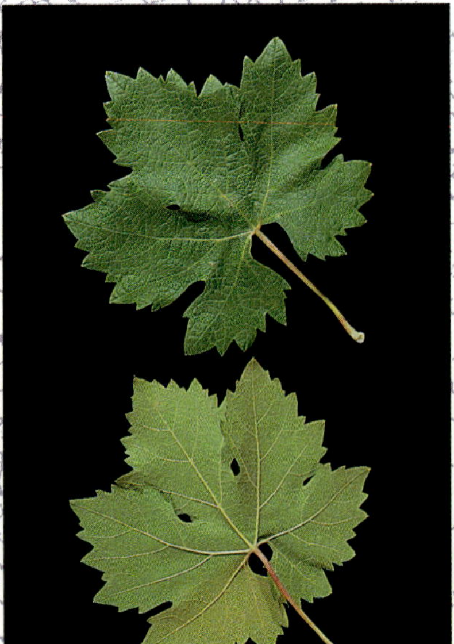

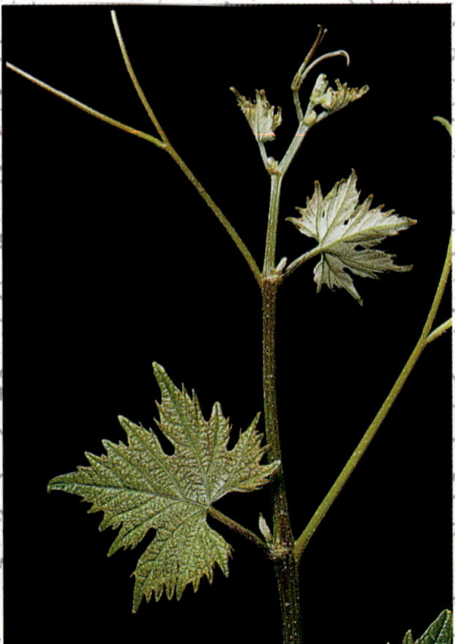

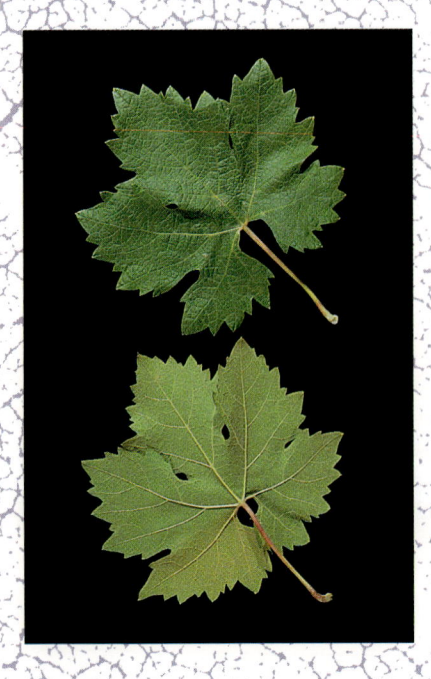

Zinfandel

Zinfandel is the Californian name for this variety and is the name under which it has been established in Australia. With the relative decline of Carignan, Zinfandel is now the second most widely planted red wine grape in California after Cabernet Sauvignon, with more than 13 800 ha in 1992 being grown successfully in all the wine grape areas. It appears to be the Italian variety Primitivo, which is grown on more than 30 000 ha in the province of Taranto. It is also grown in the Dalmatian region of Yugoslavia under the name of Plavac Veliki.

Zinfandel is a variety of moderate vigour with fairly upright shoots which mature to rather brittle canes. It has medium to large, 5-lobed leaves folding inwards with a few small tufts of hair on the lower surface. The petiolar sinus is lyre shaped, usually open but sometimes with a little overlapping of the edges. The bunches are medium to large, more or less cylindrical and compact with medium to large, round berries with a noticeable brown scar at the apex. Exposed fruit is liable to sunburn and if the vines are managed too restrictively the fruit is very liable to bunch rot.

The variety is noteworthy for the amount of second crop it produces, enough in Italy to justify a second harvest. The fruit is severely damaged by mechanical harvesting.

Zinfandel should give wines of a good bright colour and distinctive varietal character. In California it is recommended that such wines should not be blended.

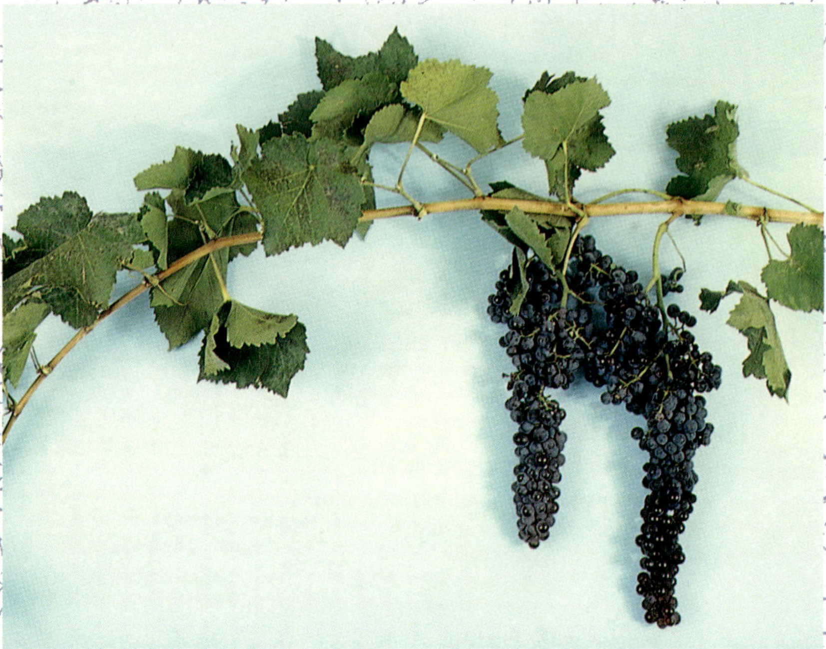

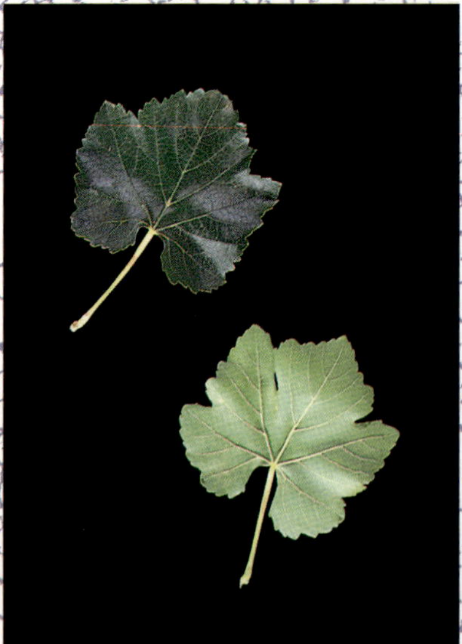

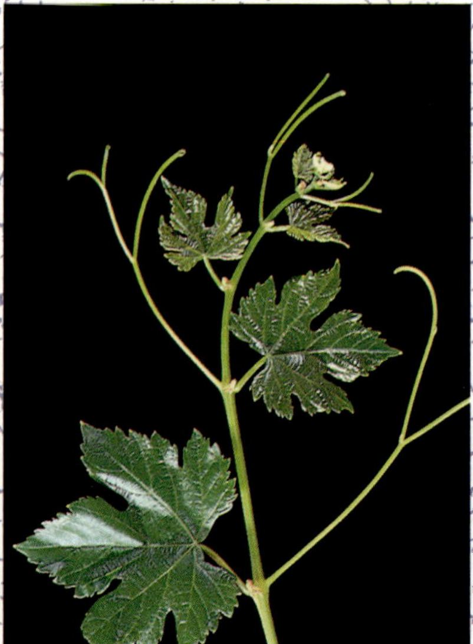

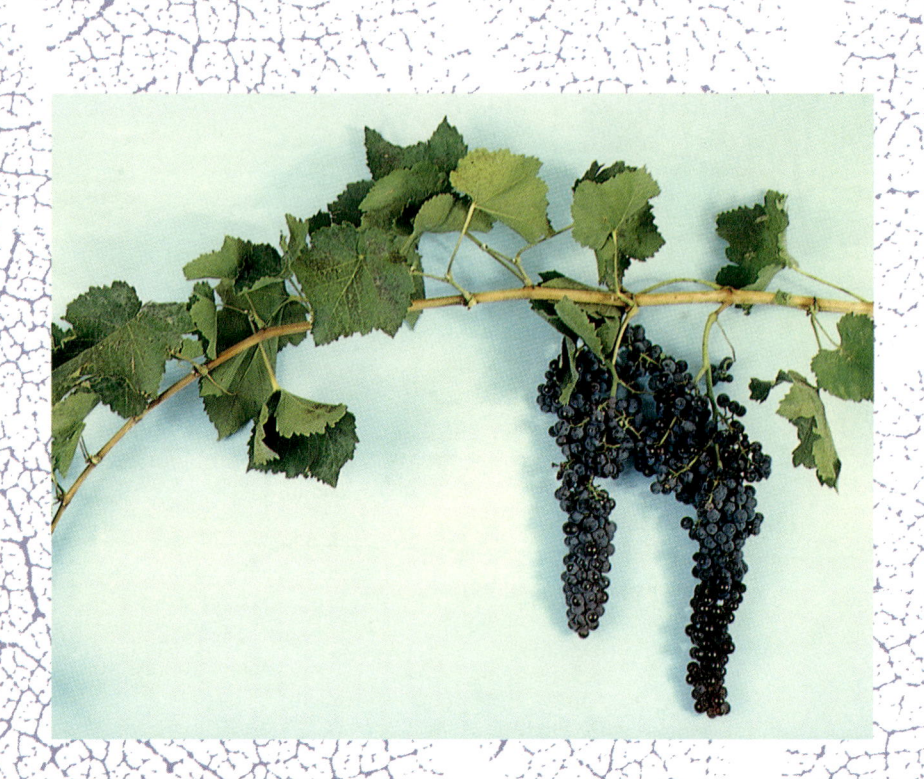

Carina

Carina is a seedless red grape bred by CSIRO Plant Industry, Horticulture Unit at its Merbein laboratory. It is a cross of Shiraz × Sultana made in 1964 and released in 1975 as a replacement for the drying variety Zante Currant. It is much less susceptible to rain damage at harvest than Zante Currant but does require the use of the same growth regulator setting sprays. It is steadily replacing the Zante for the production of dried currants and has been included in this book of wine varieties because many tonnes of Carina have been diverted to red wine making in recent years.

Carina is a vigorous variety which can produce high yields. The shoot tips are open and the young leaves are a clear green. The adult leaves are mid-green, flat, smooth and hairless. They are entire to 3-lobed with a U-shaped petiolar sinus. The teeth are broad and angular. The bunches are long, narrow, loose and cylindrical. The flowers are female. The small, round black berries are seedless and rain resistant.

The variety ripens early in mid-February at Merbein, at approximately the same time as Sultana. Carina seems to have found a use for wine making because of its earliness and the fact that it produces more deeply coloured wines than Pinot Noir in the same situation. It has also been used in the production of 'blush' sparkling wine styles.

There were 326 ha of Carina in Australia in 1996.

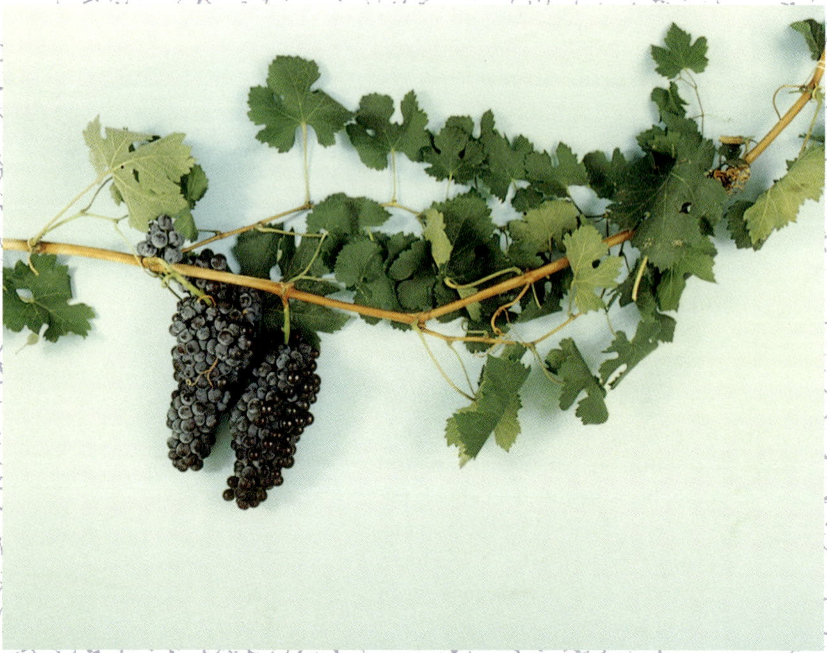

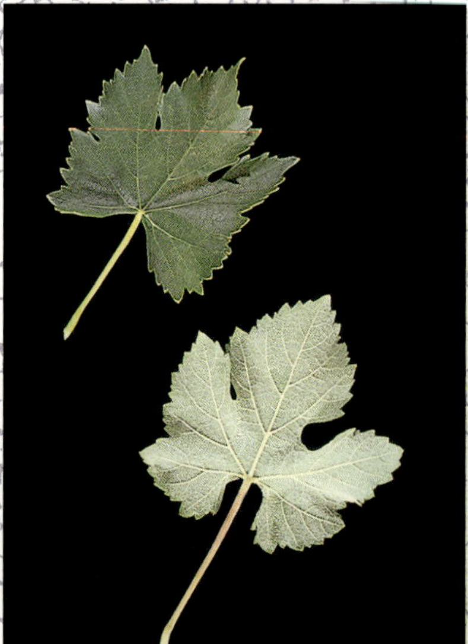

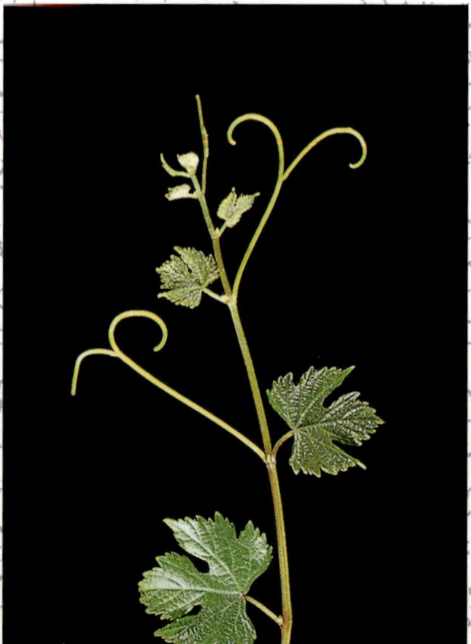

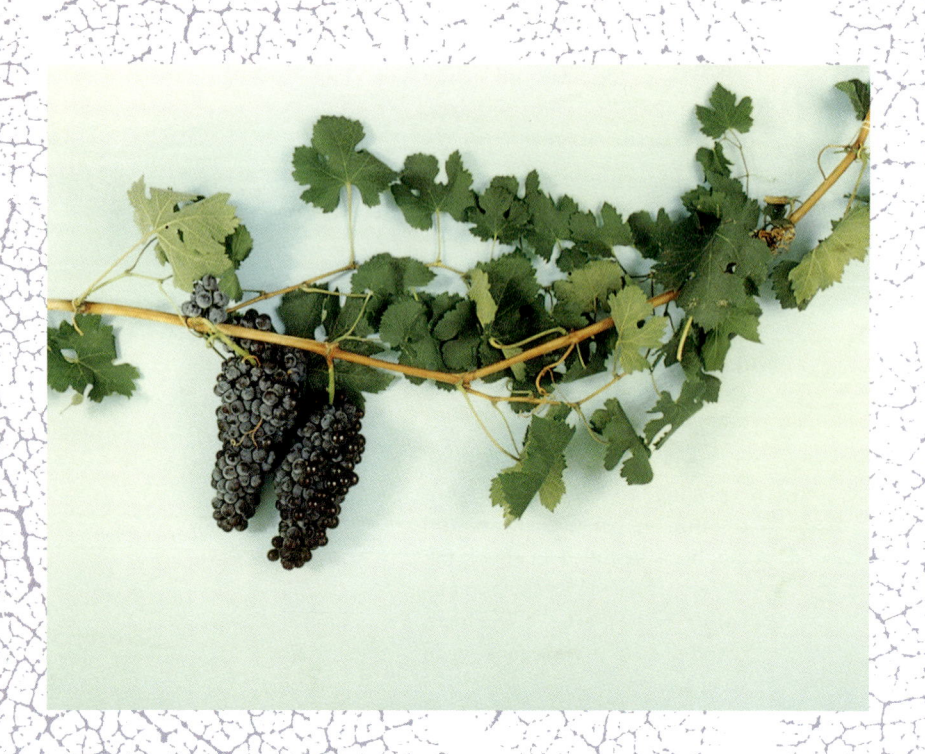

Nebbiolo

Nebbiolo is a red wine grape variety from the Piedmont region of northern Italy around Turin. Two famous wines made from the variety are Barolo and Barbaresco, named after villages in the Monferrato hills near Alba. Further north in the Novaro region the variety is known as Spanna. There are only about 5000 ha of Nebbiolo in Italy, but its reputation far exceeds its volume. There are small plantings of the variety in Switzerland, Uruguay, Argentina and California where there were 80 ha in 1997. In Australia, 17 vineyards listed plantings of Nebbiolo in 1998 in the *Australian and New Zealand Wine Industry Directory*, but the total planted area is very small.

Nebbiolo is a vigorous variety but produces only average yields. The opening buds are covered with dense hair and are greenish white in colour with red piping. The young leaves are light green with some bronzing. The adult leaves are 5-lobed and deep green with some blistering on the surface. The lower side of the leaf is covered with erect hairs. The petiolar sinus is V-shaped and sometimes closed. The bunches are medium in size, conical in shape with a prominent shoulder and reasonably compact.

The berries are medium, round and black. Maturity is late, often being harvested in November in northern Italy.

Wines made from Nebbiolo in the most favourable situations in Piedmont are high in alcohol, acid and tannin, and are traditionally aged for several years in large oak casks. This is followed by further years of bottle ageing to develop intense flavours of roses, raspberries and violets. Less favourable sites can produce softer wines with lower tannin levels. It is remarkable that a variety which produces some of the best red wines in the world has not found a place in the new world growing regions. Finding a suitable microclimate for Nebbiolo in Australia could make an interesting challenge for some aspiring young winemaker.

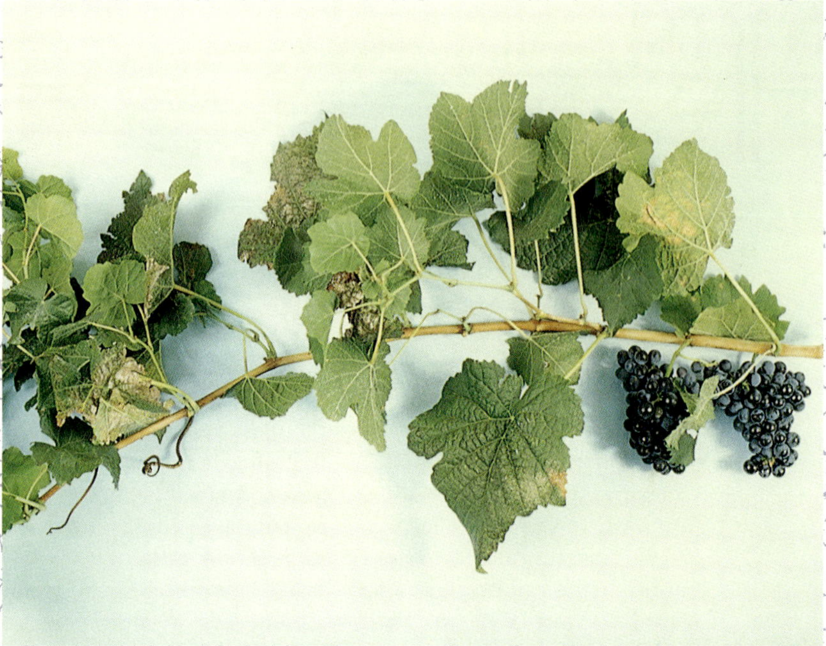

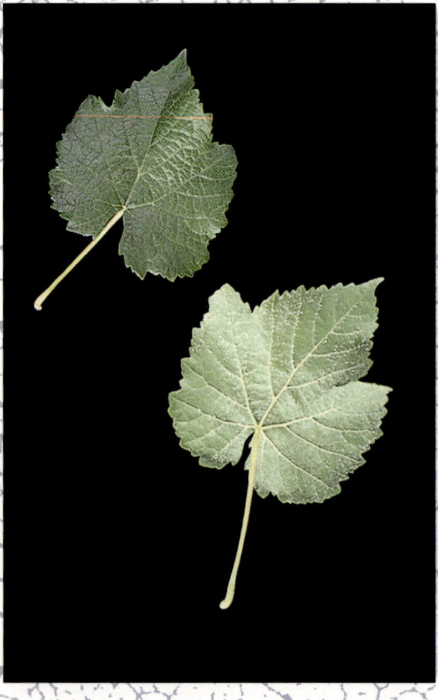

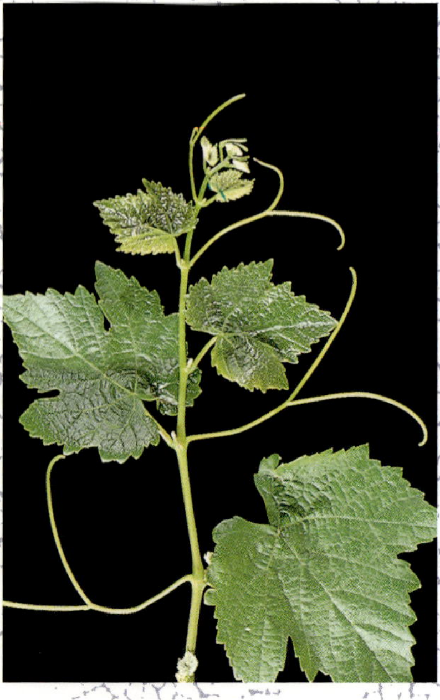

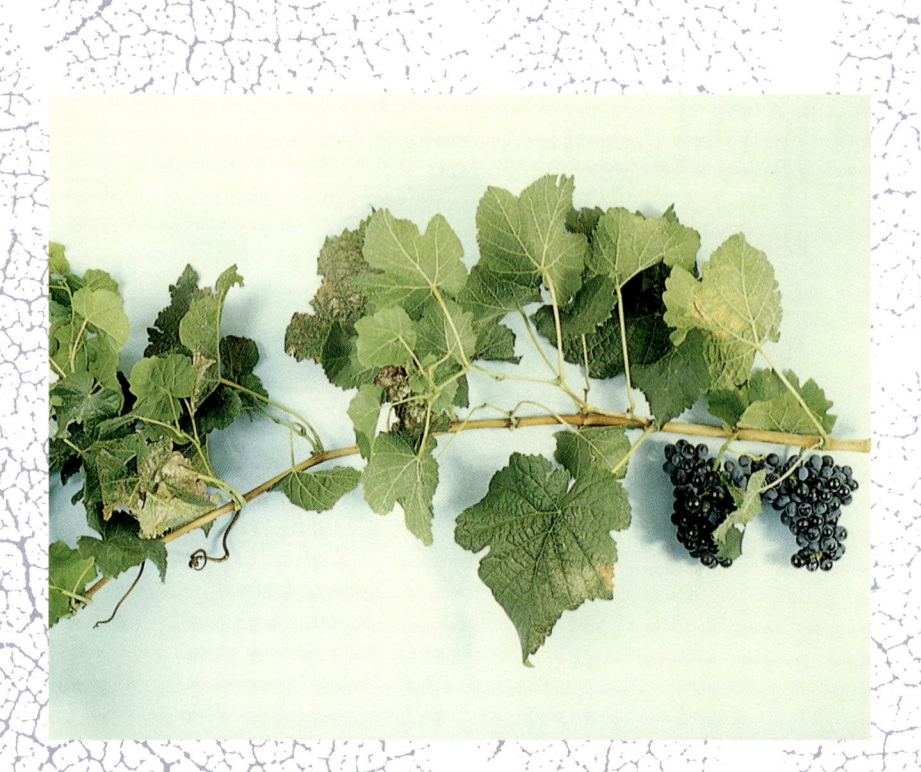

Petit Verdot

Petit Verdot is a minor grape variety from the Bordeaux region of France. It has long been part of the varietal mix in Bordeaux but probably has never exceeded 5% of the total planting. The area planted in France decreased from 685 ha in 1958 to 338 ha in 1988, but has undergone a minor revival since then to reach 380 ha in 1994. It is grown mainly on the left bank of the Gironde from Margaux to Saint-Estephe, and produces a high quality wine. It is the last variety to be harvested in the Medoc and the wines produced are well coloured, have high acid and tannin levels and age well.

In Argentina two varieties are grown under the name Verdot: one is a clone of Malbec, the other is the true Petit Verdot. It is also planted in Chile and there were 77 ha of Petit Verdot in California in 1995 with 20 ha still not in bearing. In 1998 more than 40 wineries in Australia registered Petit Verdot as a variety in the *Australian and New Zealand Wine Industry Directory*, but the total area planted was less than 100 ha. However, recently there has been increasing interest in the variety with some excellent wines being produced.

Petit Verdot is a vigorous variety which appears to have resistance to moulds. It produces high yields but the shoots tend to be long and fragile. It has been reported as having fruit-setting problems in France and the first clone introduced into Australia from California had problems with inflorescence abscission. Further clonal selection appears to have solved this problem.

The leaves are 3-lobed with the centre lobe often elongated. The leaf is dark green in colour with some blistering between the minor veins. There is some undulation between the main veins near the petiole and the petiolar sinus is a narrow lyre often with a tooth on one side. The superior sinuses are concave with closed sides and sometimes with a tooth at the base. There are medium density prostrate hairs on the lower side of the leaf. The tendrils are long, yellow and fleshy. The bunches are small to medium in size with small, round black berries.

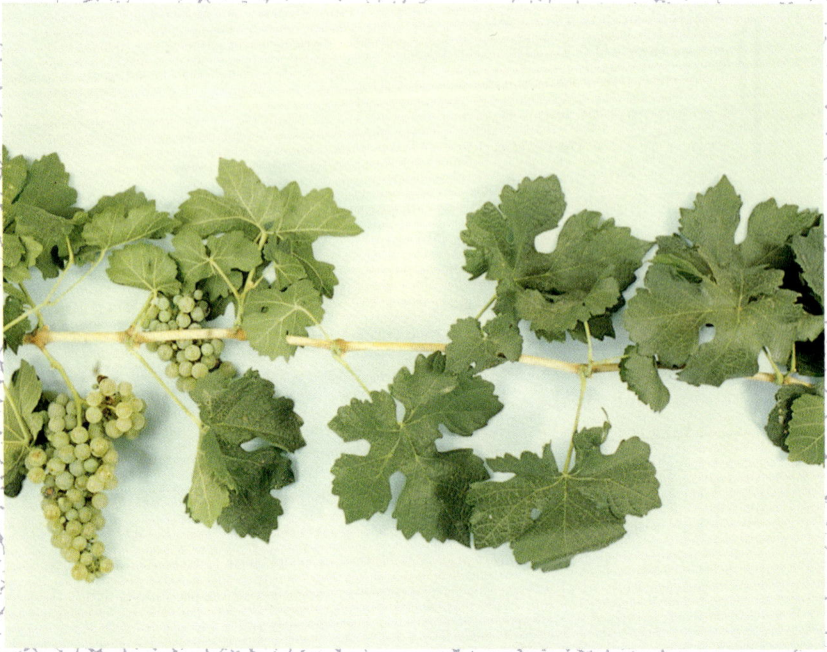
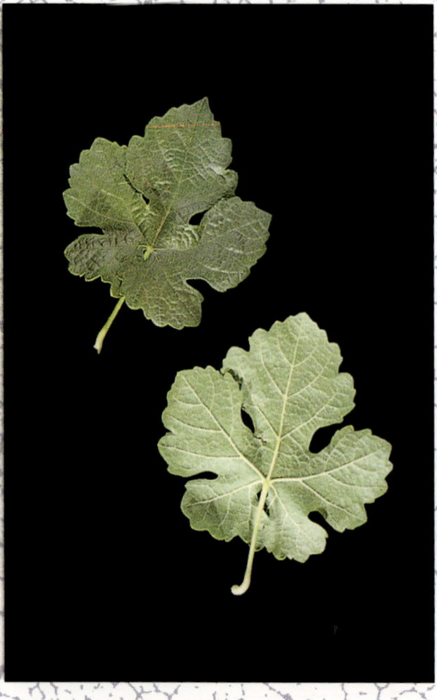
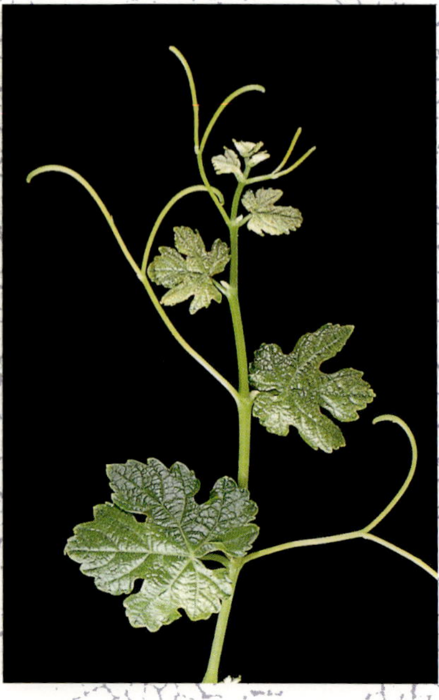

Roussanne

Roussanne is a minor white wine variety from the Rhone Valley of southern France, where it is often overshadowed by its more widely planted counterpart Marsanne with which it is often blended. There were 120 ha planted in France in 1990, mainly in the department of Herault and Vaucluse. It is a permitted addition in several regional appellations in the south of France, including Chateauneuf-du-Pape. It is also grown in the Province of Lucca in Italy. Five vineyards in Australia listed Roussanne amongst their plantings in the 1998 *Australian and New Zealand Wine Industry Directory*, but the overall area is still quite small.

It is not a vigorous variety, but it produces long shoots that can be blown off by strong winds. Yields are lower than for Marsanne, and the variety is sensitive to powdery mildew and to grey rot. It has dense prostrate hairs on the tip of the young shoots, and the young leaves are indented and covered with soft hairs. The adult leaves are also distinctly 5-lobed, large and undulating. The teeth are short compared with their width at the base. The petiolar sinus is lyre shaped and often overlapping. There are sparse prostrate hairs on the lower side of the leaf.

The bunches are small to medium, conical and often russet coloured. The berries are small, round and white. Maturity is late and this can increase problems with mould.

Wine produced from Roussanne has been described as fine and complex, with honey, floral and apricot flavours. The wines age well and develop in the bottle. It is said to add elegance and aroma to the wines of Chateauneuf-du-Pape.

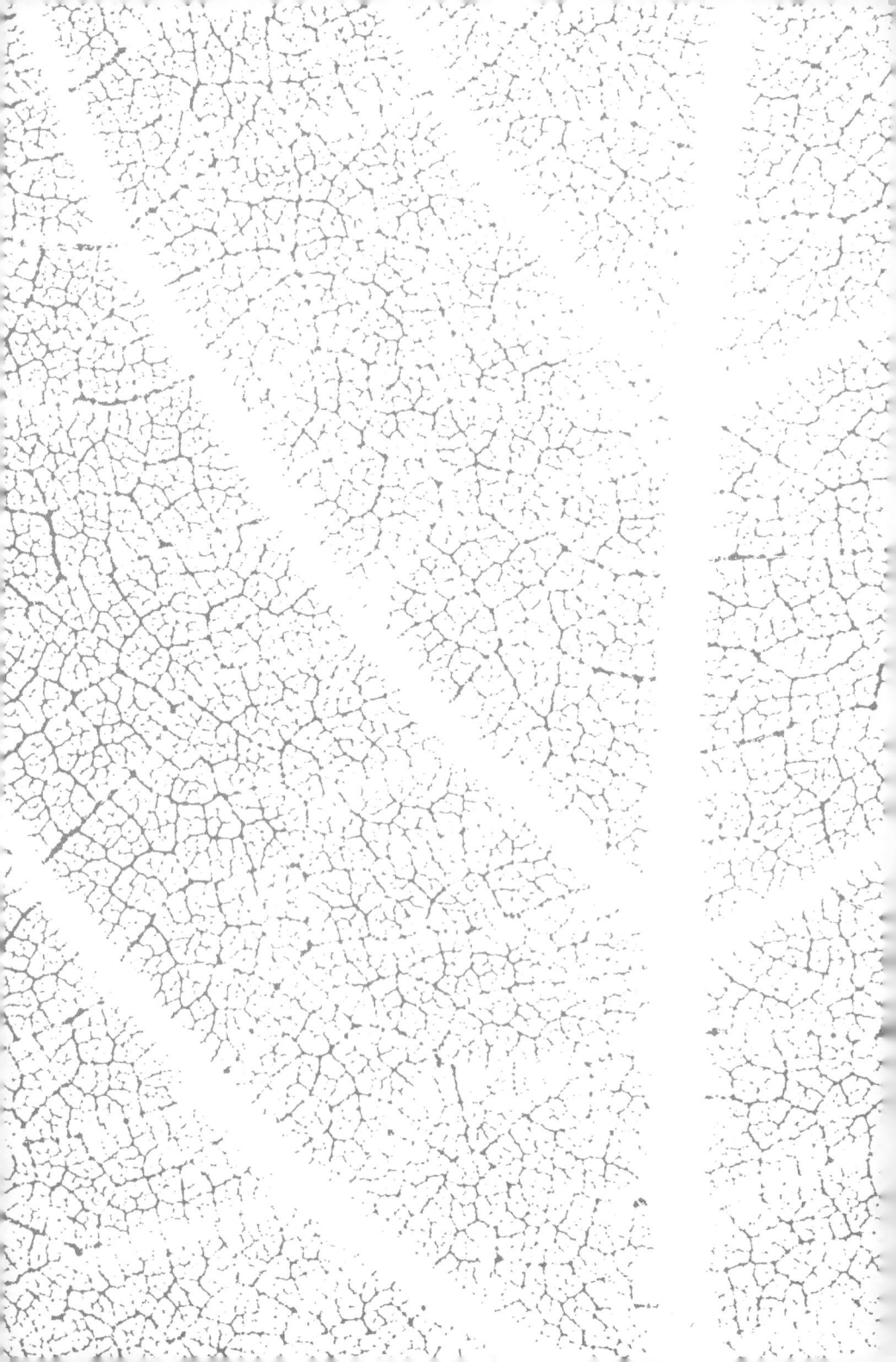

Synonyms

Grape name	Synonym (some invalid)
Aleatico	
Alvarelhao	
Barbera	
Bastardo	Trousseau (France), Cabernet Gros (South Australia)
Bianco d'Alessano	
Biancone	Pagadebiti (Corsica), Green Doradillo (Australia), Late Doradillo (Australia), White Grenache (Australia), Grenache Blanc Productif (France)
Bonvedro	Cuatendra (Spain)
Bourboulenc	Grosse Clairette, Clairette Doree
Cabernet Franc	
Cabernet Sauvignon	
Calitor	Pride of Australia (Australia)
Cañocazo	Palomino (Australia), Pedro (Australia)
Carignan	Carignane (USA)
Chambourcin	
Chardonnay	Pinot Chardonnay, Pinot Blanc Chardonnay
Chasselas	Golden Chasselas, Chasselas Dore
Chenin Blanc	Chenin (France), Pineau de la Loire (France), Steen (South Africa)
Cinsaut	Cinq-saou, Cinsault, Oeillade (France and Australia), Ulliade (Australia), Hermitage (South Africa), Ottavianello (Italy), Black Malvoisie (USA) Blue Imperial (North East Victoria, Australia), Black Prince (Western Victoria, Australia)

Grape name	Synonym
Clairette	Blanquette (France and New South Wales, Australia)
Colombard	
Crouchen	Sable Blanc (France), Riesling (South Africa), Clare Riesling (Australia)
Dolcetto	
Doradillo	
Dourado	Gallego Dourado (Portugal), Pedro (Australia), Rutherglen Pedro (Australia)
Durif	Pinot de l'Ermitage (France), Syrah Forchue (France)
Emerald Riesling	
Farana	Planta Pedralba (Spain), Mayorquin (France), Beldi (Tunisia), Damaschino (Sicily)
Fetyaska	
Flora	
Folle Blanche	
Furmint	
Gamay	
Gouais	
Graciano	Morrastel (France)
Grec rose	Wantage (Australia), Penarouch (Australia), Bermestia (Australia)
Grenache	Garnacha (Spain), Cannonao (Sardinia), Granaccia (Sicily), Alicante (Sicily)
Harslevelu	
Jacquez	LeNoir (USA), Troya (Australia)
Malbec	Cot (France), Malbeck (France), Auxerrois (France)
Mammolo	
Marsanne	Ermitage (Switzerland)
Mataro	Monastrell (Spain), Morastell (Spain), Balzac (Cognac, France and Victoria, Australia), Mourvedre (Provence, France), Esparte (Western Victoria, Australia)

Grape name	Synonym
Mauzac	
Melon	Muscadet (France)
Merlot	
Meunier	Pinot Meunier, Schwarzriesling (Germany), Miller's Burgundy (Australia)
Monbadon	Burger (USA)
Mondeuse	Refosco (Italy)
Montils	Aucarot (Australia), Aucerot (Australia)
Moschata Paradisa	
Muller-Thurgau	Riesling x Sylvaner (Switzerland), Rivaner (Luxembourg)
Muscadelle	Tokay (Australia), Sauvignon Vert (California)
Muscat a Petits Grains	Moscatel Menudo (Spain), Muscat de Frontignan (France), Muscat d'Alsace (France), Moscato d'Asti (Italy), Moscato di Canelli (Italy), Frontignac (France)
Muscat Gordo Blanco	Muscat of Alexandria, Moscatel de Malaga (Spain), Muscat de Setubal (Portugal), Zibibbo (Italy), Hanepoot (South Africa)
Muscat Ottonel	
Ondenc	Piquepout de Moissac (Armagnac, France), Blanquette (Bordeaux, France), Blanc Selection Carriere (Cognac, France), Dourec (Pyrenees), Plant de Gaillac (Pyrenees), Irvine's White (Victoria, Australia)
Orange Muscat	Muscat Fleur d'Oranger (France), Muscat Primavis, Muscat de Jesus
Palomino	White French (South Africa), Listan (France), Paulo (Australia)
Pedro Ximenez	
Peloursin	
Pinot Blanc	
Pinot gris	Rulander (Germany), Szurkebarat (Hungary)
Pinot Noir	
Rkaziteli	Gratiesti (Moldavia), Sonnenkuste (Bulgaria)

Grape name	Synonym
Riesling	Rhine Riesling (Autralia), Riesling Renano (Italy), Rheinriesling (Austria), Rajnski Rizling (Yugoslavia), Rajnai Rizling (Hungary), White Riesling (California)
Rubired	
Ruby Cabernet	
Sangiovese	
Sauvignon Blanc	
Semillon	Greengrape (South Africa), Hunter River Riesling (Australia), Barnawartha Pinot (Australia)
Shiraz	Syrah (France), Hermitage (Australia)
Siegerrebe	
Solvorino	
Souzao	Vinhao (Portugal), Souson (Spain)
Sultana	Thompson Seedless (USA), Sultanine (Western Europe), Sultanina (Western Europe)
Sylvaner	Silvaner (Germany)
Taminga	
Tarrango	
Tempranillo	Tinta Fino (Spain), Cencibel (Spain), Tinta Roriz (Portugal), Negretto (Italy), Valdepenas (USA)
Terret Noir	
Tinta Amarella	
Tocai Friulano	Sauvignonasse (France)
Touriga	
Traminer	Highly aromatic clones referred to as Gewurztraminer
Trebbiano	Trebbiano Toscano, Ugni Blanc (France), Saint Emilion (Cognac, France), White Shiraz (Australia), White Hermitage (Australia), Clairette Ronde (France, Italy)
Valdiguie	Napa Gamay (Australia, USA)
Verdelho	Gouveio (Portugal)
Viognier	
Zinfandel	Primitivo (Italy), Plavac Veliki (Yugoslavia)

Variety Collection

CSIRO Division of Plant Industry (Horticulture Section)
Merbein, Victoria
Grapevine Variety Collection
As at 9 October 1995

101-14 Millardet	554-5 Couderc	Arinarnoa
106-8 Millardet	Seedlings	Arinto
107-11 Millardet	57 Richter	Arkansas 1105
110 Richter	62-66 Couderc	Arnsburger
1103 Paulsen	99 Richter	Arriloba
1202 Couderc	Abouriou	Arrouya
125 AA Kober	Agadaj	Askari
125-1 Millardet	Agawam	Aspiran
128 Seibel	Agostenga	Aspiran Bouschet
13-5 EVEX	Albarin	Aubun
140 Ruggeri	Albillo	Aurelia
157-11 Couderc	Alden	Aurora
160-19 Couderc	Aleatico	Auxerrois
161-49 Couderc	Aledo	Baco 22A
1613 Couderc	Alicante Bouschet	Bakator
1616 Couderc	Aligote	Balluti
188-04 Castel	Alvarelhao	Banatski Muscat
19-52 Millardet	Anab-e-Shahi	Barbera
219-A Millardet	Antao Vaz	Baresana
301-A Millardet	Antigona	Barlinka
31 Richter	Aramon	Baroque
3306 Couderc	Aramon Bouschet	Barry
3309 Couderc	Aramon Gris	Bastardo
333 EM Foex	Arbane	Baufrac
34 EM Foex	Arbois	Baxter's Sherry
41 B Millardet	ARG 1	Beauty Seedless
420 A Millardet	ARG 9	Bebas Blanco
5 BB Kober	Argant	Beclan

Bells Seedling	Burgrave	Chancellor
Beogradska Besemena	Cabernet Franc	Chaouch
Beogradska Rana	Cabernet Franc	Chardonelle
Bequignol	Francese	Chardonnay
Bergonia	Cabernet Sanzey	Chardonnay Muscat
Best's A	Cabernet Sauvignon	Chasselas Dore
Best's B	Caino	Chenin Blanc
Best's C	Calagrano	China Seedling
Best's K	Calitor	Christmas Rose
Best's L	Calmeria	Cinsaut
Best's M	Calzin	Cionoria
Best's N	Campbell's Early	Clairette
Best's Q	Canada Muscat	Clairette Poinseau
Best's T	Canadice	Clairette Rose
Best's No.22	Canner	Classelas Lacinie
Best's R15V92	Cannon Hall Muscat	Clersole Logine
Best's R1V8	Canocazo	Colombard
Best's R2V73	Cape Currant	Comtessa
Bianca	Captivator	Concord
Bianco d'Alessano	Cardinal	Constantia
Biancolella	Carignan	Corbeau
Biancone	Carina	Cornifesto
Bicane	Carolina Blackrose	Cortese
Black Chilean	Cascade	Corvina Veronese
Black Frontigan	Castor	Corvinon
Black Hamburg	Cataratto	Cosmo 10A
Black Malaga	Catawba	Cosmo 2B
Black Monukka	Cayuga White	Couderc Noir
Black Prince	Cegled Szepe	Crimson Seedless
Blackrose	Centennial	Crinto
Blush Seedless	Centennial Seedless	Criolla Negra
Boal	Cereza	Criolla San Juanina
Bobal	Cesanese	Croattina
Boglarka	CG 1481	Crouchen
Bombino Nero	CG 1730	Crujidera
Bonarda	CG 183	Danlas
Bonvedro	CG 26879	Danugue
Borner	CG 351	Dawn Seedless
Bourboulenc	CG 38049	De Chaunac
Brachetto	CG 4320	Delaware
Bruce's Sport	Chali Sar	Delight
Buckland Sweetwater	Chambourcin	Delizia di Vaprio

Demir Kapija	Favorit	GM 647-2
Diamond	Feher Szagos	GM 6495-1
Dizmar	Fer	GM 6495-3
Djandjal Kara	Fer Servadou	GM 7116-2
Dog Ridge	Fercal	GM 723-1
Dolcetto	Ferdinand de Lesseps	Godello
Doradillo	Fernao Pires	Goethe
Dourado	Fetyaska	Gold
DRX 55	Fiano	Golden Champion
Duchess of Buccleuch	Fiesta	Golden Muscat
Dunkelfelder	Flame Seedless	Golden Queen
Durif	Flame Tokay	Gordana 2
Dutchess	Flora	Gordana 3
Early Muscat	Frankenthal	Gouais
Early Niabell	Fredonia	Goyura
Egiodola	Freedom	Graciano
Ehrenfelser	Freisa	Graham's Black
Elvira	Freisamer	Grand Noir
Emerald Riesling	Fresno 18-149	Grec Rose
Emerald Seedless	Fresno 27-31	Green Hungarian
Emperatriz	Fresno 32-145	Green Veltliner
Emperor	Fresno 58-93	Grenache
Etraire de la Dui	Fresno G4-74	Grey Riesling
Exotic	Fresno Seedless	Grignolino
Ezerfurtu	Fuji Muscat	Grillo
Ezerjo	Furmint	Grocanica
False Alicante Provencial	Gamay	Grolleau
	Gamay de Bouze	Gropello Gentile
False Black Malvoisie	Gamay Freaux	Gros Colman
False Calitor	Garonnet	Gros Manseng Blanc
False Clairette Blanche	Gascon	Gros Meslier
False Clairette de Limou	Geisler's Glory #1	Gros Syrah
	Geisler's Glory #2	Gros Vert
False Concord	Gf V31-17-115	Grosse Blaue
False Keknyelu	Gf V3125	Gueche Noir
False Mataro	Glenora	Gutadel x Sylvaner
False Muscadelle du Bordelais	GM 316-57	Gyongyrizling
	GM 318-57	Harmony
False Nocera	GM 322-58	Harslevelu
False Westbury	GM 324-58	Hawson's Seedless Muscat
Fantasy Seedless	GM 6414-11	
Farana	GM 643-16	Hawson-Geisler Selection

Helena	K 51-32 Lider	Malvasia Istriana
Henab Turki	K 51-40 Lider	Malvasia Rioja
Herbemont	Kadarka	Malvoise
Herbert	Kali Sahebi	Mammolo
Himrod	Kandahar	Manik Chaman
Hunisa	Katta Kourgan	Mantey
Husseine	Kavadarski Drenak	Mantua de Pilas
IAC 313	Keknyelu	Marechal Foch
Illinois 271-1	Kerner	Marocain Noir
Illinois 547-3	Kiralyleanyka	Marsanne
Incrocio Bruni	Kishmishi	Marzemino
Ingram's Hardy Prolific	Kladovaska Bela	Mataro
	Koshu Sanjaku	Mauzac
Insolia	Kyoho	Mavro Naussis
Iona	La Reina	Mavro Romeico
Irsay Oliver	Lady Downes Seedling	Mavronemeas
Isabella	Lady Patricia	Medea
Italia	Lagrein	Melon
J 17-09 Lider	Lakhegyi Mezes	Menavacca
J 17-48 Lider	Lambrusco	Merbein Seedless
J 17-53 Lider	Le Bayard	Merlot
J 17-58 Lider .	Leanoy	Merlot Blanc
J 17-69 Lider	Len de L'El	Metallica Cape
Jacquez	Leon Millot	Meunier
Jaen	Lignan Blanc	Michurinets
JS23-416	Liliorila	Miguel de Arco
July Muscat	Limberger	Mills
Jurancon Noir	Lival	Minnella Bianca
K 5-2 Lider	LN 33	Mission
K 5-4 Lider	Loose Perlette	Mission Seedling
K 5-20 Lider	Loureiro	Molinara
K 48-15 Lider	Luglienga	Monastrell
K 48-29B Lider	Macabeo	Monbadon
K 48-35 Lider	Madeleine Angevine	Mondeuse
K 48-38 Lider	Madio Seedling	Montepulciano
K 48-43 Lider	Madresfield Court	Montils
K 48-45 Lider	Muscat	Morio Muskat
K 48-48 Lider	Malbec	Morrastel Bouschet
K 48-69 Lider	Malta Seedless	Moschata Paradisa
K 49-56 Lider	Malvasia Bianca	Mourisco Branco
K 51-1 Lider	Malvasia di Candia	Mrs Pince's Muscat
K 51-11 Lider	Malvasia Istria	Mtsvane

Muller Thurgau
Muscadelle
Muscat a Petits Grains
  Blanc
Muscat a Petits Grains
  Rouge
Muscat Bailey A
Muscat de St. Vallier
Muscat Flame
Muscat Gordo Blanco
Muscat Hamburg
Muscat Ottonel
Muscat Seedless
Naza Valenciana
Nebbiolo
Nebbiolo Chiavenas
Negra Molle
Negrette
Negro Amaro
Negru Virtos
New York Muscat
Nimrang
Noir Hatif de
  Marseille
Nyora
Odola
Ohanez
Olmo H34-15
Ondenc
Ontario
Opuzensia Rana
Orange Muscat
Orion (GA-58-30)
Ortrugo
Palomino
Pannonia Gold
Parellada
Pedro Ximenez
Peloursin
Perdea
Perle
Perle de Csaba

Perlette
Perlon
Pescatore's No.10
Petit Bouschet
Petit Maseng
Petit Meslier
Petit Verdot
Peverella
Picolit
Pignoletto
Pinot Blanc
Pinot Gris
Pinot Noir
Pinot Precoce
Pione
Piquepoul
Piros Slanka
Poerinha
Pollux
Ponce Precice
Portugais Bleu
Precoce de Malingre
Preto Martinho
Primiera
Primitivo di Gioia
Prokupac
Purple Cornichon
Putzscheere
Queen
Queen of the Vineyard
Rabaner
Rabigato
Rabioso Piave
Radmilovaski Muskat
Raffiat de Moncade
Raisin de Palestine
Ralph's Red
Ramsey
Rauschling
Red Globe
Red Malaga
Red May

Red Prince
Red Sultana
Red Veltliner
Refosco Nostrano
Refosco Pedunclo
Rosso
Reichensteiner
RF 48
Ribier
Ribol
Riesling
Riesling Italico
Riparia Gloire de
  Montpellier
Rish Baba
Rkaziteli
Rolle
Romulus
Rondinella
Rose of Peru
Rossignola
Roter Zierfandler
Rotgipfler
Rounce's Early
Roupeiro Cachudo
Roussanne
Royal Ascot
Royalty
Rubired
Ruby Cabernet
Ruby Seedless
Rufete
Rupestris du Lot
Russian Seedless
Sabalkanskoi
Saint Laurent
Saint Macaire
Sangiovese
Santa Paula
Saperavi
Sauvignon Blanc
Sauvignon Rose

Scarlet	Sylvaner	Verdal
Scheurebe	Tafifi	Verdejo Blanco
Schonburger	Takao	Verdelet
Schuyler	Taminga	Verdelho
Schwarzmann	Tandanya	Verdicchio
Sciacarello	Tannat	Verduzzo Friulano
Seedless Emperor	Tarrango	Vermentino
Seedless Muscat	Teleki 5A	Vespolina
Seibel 10868	Teleki 5C	Victoria Blanc
Seibel 14664	Teleki 8B	Villa Nueva
Semebat	Teleki A	Villard Blanc
Semillon	Teleki B	Villard Blanc x Perlette
Semillon Rose	Teleki C	Villard Noir
Semis de Goes	Temperano	Vincent
Seneca	Tempranillo	Vinered
Sereksiya Black	Teneron	Viognier
Severnyi	Teroldego	Viosinho
Seyval	Terret Noir	Vital
Shakar Angur	Thomuscat	*Vitis aestivalis*
Shirana 6	Tinta Amarella	*Vitis amurensis*
Shirana 67	Tinta Cao	*Vitis berlandieri*
Shirana 68	Tinta Carvalha	*Vitis candicans*
Shiraz	Tinto	*Vitis caribaea*
Shtur Angur	Tocai Friulano	*Vitis champini*
Siderites	Touriga	*Vitis cinerea*
Siegerrebe	Trajadura	*Vitis cordifolia*
Smederevka	Traminer	*Vitis girdiana*
SO 4	Traminer x Riesling 25/4	*Vitis longii*
Solvorino		*Vitis riparia*
Sori	Trebbiano	*Vitis rotundifolia*
Souzao	Trieste 4X	*Vitis rupestris*
Suffolk Red	Troyen	Waltham Cross
Sultana	Tulillah	White Muscat
Sultana Monococco	Ughetta	White Vernaccio
Sultana Moschata	Urbana	Wortley Hall
Sultana Nera	Uva di Troia	Xarello
Sultanina Agrio	V.rupestrix R65-36	Zalagyongye
Sultanina Rose	Valdiguie	Zante Currant
Sumoll	Varousset	Zinfandel
SV 39-639	Veeport	
Sweetwater	Venus	

Glossary

Appellation
A system by which some European countries seek to maintain quality and product image for their wines. For example, the system of Appellation d'Origine Controllée in France specifies which areas of land may be used for grape growing, which cultivars may be used, the maximum yield and the minimum alcohol content of the wines. Australia does not have this type of legislation.

Black Spot
A serious fungal disease of grapevines, particularly the cultivar Sultana and some tablegrapes; the causal organism is *Elsinoe ampelina*. Characteristic symptoms are deformed shoots with black cankers and distorted leaves bearing small purple-black spots.

Bloom
A natural layer of wax on the surface of the berry which prevents the loss of water from the fruit.

Botrytis
Botrytis bunch rot is a fungal disease of grapevines throughout the world. The causal organism is *Botrytis cinerea*. It causes loss of bunches through rotting of the rachis, loss of juice and rotting of the fruit. Under certain conditions the growth of *botrytis cinerea* on ripe bunches can take on a form known as 'noble rot'. This desiccates the berries and enhances the sweetness and flavour of the juice.

Bullate
A term describing the appearance of the leaf surface. Bullate leaves have tiny bulges emerging from between the smallest subdivisions of the veinlets which are only a few millimetres in diameter (e.g. Pinot Noir).

Bunch rot
See Botrytis.

Cane
A mature shoot of a grapevine.

Cane pruning
A pruning system where long rods or canes of 10 to 15 buds are retained as fruiting units for the following season's crop.

Chimaera
A mixture of normal and mutant tissue. In a periclinical chimaera the plant is composed of a mutant 'skin' enclosing a normal interior. An example of periclinical chimaera is seen in the cultivars Pinot Meunier and Pinot Noir.

Clone
A group of individual plants propagated asexually from a single ancestor.

Downy mildew
A disease of grapevines caused by the fungus *Plasmopara viticola* which attacks all of the green parts of the vine. Severe infections can cause defoliation leading to reduced sugar accumulation in the berries and the loss of winter hardiness by latent buds. The characteristic lesions are yellow 'oil' spots on the leaves.

Fortified wine
Wine which has received an addition of grape spirit or brandy.

Gall
Crown gall is a disease caused by a soil bacterium *Agrobacterium tumefaciens*. It is a systemic disease which produces characteristic galls on the stems of grapevines. Serious infections can cause reduced yield and vigor.

Glabrous
A term applied to leaves which are without hairs or down.

Hen and chicken
A disorder of the grapevine where both mature seeded berries and immature seedless berries may be retained by a bunch.

Hermaphrodite
Having both male and female parts on the same flower. Most commercially important *Vitis* cultivars are hermaphrodite whilst some rootstocks are female and some are male.

Internode
The portion of a cane or shoot between two consecutive nodes.

Midrib
Main vein down the centre of a leaf.

Noble rot
See Botrytis.

Node
Thickened part of the shoot where the leaf arises and where buds are formed.

Oblate
Flattened at the poles as in the berries of *V. riparia*.

Obtuse
Blunt, not sharp or pointed.

Oidium
Powdery mildew, or oidium is a major fungal disease of grapevines. The causal organism is *Uncinula necator*. The symptoms are ash-grey to white powdery growth on both the upper and lower surface of leaves or on bunches. Severe disease delays the maturity of berries, retards their growth and causes them to split.

Orbicular
Spherical or circular.

Peduncle
The stalk by which the bunch or cluster is attached to the vine.

Petiole
The stalk of a leaf.

pH
A measure of the acidity or alkalinity of a solution. The pH value is the logarithm of the reciprocal of the hydrogen ion concentration. pH 7 is neutral, values below 7 are acid and above 7 alkaline.

Phylloxera
A hemipterous insect, *Daktulosphaira vitifolii* of North American origin which attacks the leaves and particularly the roots of grapevines. A small yellow aphid-like insect, it is regarded as the world's worst grape pest.

Plane
Flat.

Powdery mildew
See Oidium.

Raisin
Dried grape.

Root knot nematode
Species of the genus Meloidogyne. A serious soil pest which attacks grapevine roots.

Sinus
The opening between the lobes in the blade of the leaf.

Sport
Bud mutant.

Spur pruning
A pruning system where only short 2–3 node spurs are retained as fruiting wood, usually on established permanent arms or cordons. Traditional spur pruning removes more than 70% of the potential fruit bearing wood in any one year.

Tannin
A group of organic compounds characterised by their astringent taste. They contribute to the flavour in wine and are derived from grape skins and seeds.

Teeth
The protuberances of the grape leaf margin are called 'teeth'. They can be pointed, convex, concave or hooked in shape and may vary in size.

Tendril
The slender, leafless organ by which a vine shoot attaches itself to a support.

Young leaves
Leaves which have separated from the growing tip but still have not reached maturity.

References

Alleweldt, G. and Dettweiler, E. (1992). *The genetic resources of Vitis*, 3rd edn, Institut für Rebenzuchtung Geilweilerhof, Siebeldingen.

Antcliff, A. J. (1976a). Variety identification in Australia: A French expert looks at our vines, *Australian Grapegrower and Winemaker* 153, 10–11.

Antcliff, A. J. (1976b). *Some wine grape varieties for Australia*, CSIRO, Adelaide.

Antcliff, A. J. (1979). *Major wine grape varieties of Australia*, CSIRO, Adelaide.

Antcliff, A. J. (1983). *Minor wine grape varieties of Australia*, CSIRO, Adelaide.

Busby, James. (1833). *Journal of a tour through some of the vineyards of Spain and France*, Stephen and Stokes, Sydney, (Facsimile reprint: David Ell Press, Hunters Hill, NSW, 1979).

Dettweiler, E. (1991). *Preliminary minimal descriptor list for grapevine varieties*, Institut für Rebenzuchtung Geilweilerhof, Siebeldingen.

Galet, P. (1985). *Precis d'ampélographie pratique*, Imprimerie Charles Dehan, Montpellier.

Gregory, G. R. (1988). Development and status of Australian viticulture, in *Viticulture*, Volume 1, Resources in Australia, eds B. G. Coombe and P. R. Dry, Australian Industrial Publishers, Adelaide, pp. 1–36.

Major, M. ed. (1998). *The Australian & New Zealand wine industry directory*. 16th ed. Winetitles, Marleston, S. Aust.

Viala, P. and Vermorel, V. (1909). *Ampélographie*, 7 volumes, Masson, Paris.